PENGUIN

FEARFUL S

Ian Stewart is Professor of Mat[illegible] University of Warwick, and the author of over fifty books, including the best-selling *Does God Play Dice?* (Penguin 1990) and *Game, Set and Math* (Penguin 1991). He writes the 'Mathematical Recreations' column in *Scientific American,* is mathematics consultant to *New Scientist* and often appears on radio. He is a research mathematician working on dynamical systems, aiming to develop practical applications of new ideas from pure mathematics. As a hobby he writes science fiction and keeps fish.

Martin Golubitsky is Cullen Distinguished Professor of Mathematics and Director of the Institute for Theoretical and Engineering Science at the University of Houston. He has written extensively on both how symmetries are useful when solving mathematical models and what implications these ideas have for experiments. Recently Golubitsky has concentrated his efforts on how symmetry and chaos combine to generate a new method for pattern formation, and he is co-author of *Symmetry in Chaos* (1992).

Both authors were drawn together by a common interest in the application of new mathematical ideas to scientific problems, worked together for a year in Houston between 1983 and 1984 and have been collaborating intermittently on various projects ever since.

IAN STEWART
AND MARTIN GOLUBITSKY

FEARFUL SYMMETRY

IS GOD A GEOMETER?

PENGUIN BOOKS

PENGUIN BOOKS

Published by the Penguin Group
Penguin Books Ltd, 27 Wrights Lane, London W8 5TZ, England
Penguin Books USA Inc., 375 Hudson Street, New York, New York 10014, USA
Penguin Books Australia Ltd, Ringwood, Victoria, Australia
Penguin Books Canada Ltd, 10 Alcorn Avenue, Toronto, Ontario, Canada M4V 3B2
Penguin Books (NZ) Ltd, 182–190 Wairau Road, Auckland 10, New Zealand

Penguin Books Ltd, Registered Offices: Harmondsworth, Middlesex, England

First published by Blackwell Publishers, by arrangement with Penguin Books, 1992
Published in Penguin Books 1993
1 3 5 7 9 10 8 6 4 2

Printed in England by Clays Ltd, St Ives plc

Contents

God ever geometrizes

Plato

Geometrical properties are characterized by their invariance under a group of transformations

Felix Klein

If Plato and Klein are correct, then God must be a group-theorist. Is She?

Fearful Symmetry

Figures

Preface

When we decided to write a sequel to *Does God Play Dice?*, we knew it was going to be a hard act to follow. Chaos, the subject of that book, was – and still is – one of the major growth areas of scientific research, where the frontiers of mathematics and physics collide; its applications range from the transmission of epidemics to the orbital motion of Pluto. Moreover, it rides on the back of a philosophical paradox: how can deterministic mathematical models produce random behaviour? You can get a lot of mileage out of a philosophical paradox.

So, instead of following the act, we decided to precede it. Logically (but not chronologically) *Fearful Symmetry: Is God a Geometer?* comes before its predecessor: it's what movie-makers call a prequel. Does God Minus One. It's about patterns, rather than their absence. The way we've written it, it's about the area in which *we* do research, which isn't of itself a major global movement. However, as the writing of the book progressed, it became our private picture of a global movement far vaster than chaos, one that's been around for centuries: the search for mathematical explanations of nature's regularities. Chaos, of course, is really part of that enterprise too: it extracts new regularities from apparent disorder.

Despite its subtitle, *Fearful Symmetry* isn't a theology book, or a geometry text. It's about the role of symmetry in pattern formation. A lot of books, including several classics, have been written about symmetry in mathematics, art, science, and nature. So what's new? The answer is that most of those books view symmetry as a static phenomenon: an object either has symmetry or it doesn't. If they do any mathematics they usually try to teach you about group theory, which is the mathematics that defines, organizes, and classifies symmetry. They will tell you that William Blake's celebrated 'tyger', to whom our title pays homage, still looks like a tyger when you view it in a mirror.

Group theory is given a mention here, but only as a language. Mathematically, we take a rather different line, emphasizing dynamic processes in which symmetry can either be destroyed or created. For us, the fearful symmetry of the tyger lies in the dynamics of biological development, which creates a miracle of striped perfection from a spherical egg less than a millimetre across. Fearful, indeed – but in the sense of 'awe-inspiring'. So, instead of just classifying, enumerating, and recognizing various types of symmetry in a thousand manifestations, we ask – and offer answers to – two basic questions. John Kenneth Galbraith once wrote a book called *Money: Where It Came From, Where It Went*; we tackle the same questions for symmetry. What is it, and where did it go?

Chaos demonstrates that deterministic causes can have random effects – a big surprise, even if everybody now claims they knew it all along. There's a similar surprise regarding symmetry: symmetric causes can have asymmetric effects. This *must* be a surprise, because it directly contradicts a celebrated principle formulated by the great physicist Pierre Curie. Of course, it turns out that with the right interpretation, Curie was right all along; but many of the consequences people draw from his principle are wrong, because they use the wrong interpretation.

This paradox, that symmetry can get lost between cause and effect, is called *symmetry-breaking*. In recent years scientists and mathematicians have begun to realize that it plays a major role in the formation of patterns. How can the destruction of symmetry create pattern? That's one aspect of the paradox. Part of its resolution is that the time-honoured term 'breaking' is poorly chosen. Some quirks of the human pattern-recognition system are also involved.

Among the patterns that you'll meet in this book are many that occur on a human scale, such as the stripes on Blake's tyger, corn circles, dewdrops on a spiderweb, and the pattern in which centipedes move their legs. On larger scales they range through the chain of volcanoes that makes up the Hawaiian islands, the vibrations of stars, the spiral swirl of galaxies, even the universe itself. In the opposite direction, they include the arrangement of atoms in a crystal, the shape of viruses, and the structure of subatomic particles. From the smallest scales to the largest, many of nature's patterns are a result of broken symmetry; and our aim is to open your eyes to their mathematical unity.

The interaction of symmetry with dynamics is itself a rapid growth area of research: it may not be as well known as chaos, but in some respects it's probably more useful. Humanity has always been better at exploiting patterns than it has chaos; though perhaps better at generating chaos than order. It may seem that there is an unbridgeable gulf between symmetry and chaos; but one of the most exciting prospects for future discoveries lies in their interaction. In the dynamic behaviour of systems with symmetry mathematicians have stumbled upon a natural meeting-place for order and chaos, a story that we tell in the penultimate chapter.

Traditionally, a preface is a place to thank friends and colleagues for their assistance. We could probably fill a book the size of this one with the names of all the people who have, in one way or other, contributed. We have room only to name a few. They include Mark Allin, Peter Ashwin, Fritz Busse, Pascal Chossat, Jack Cohen, Jim Collins, John David Crawford, Michael Dellnitz, Russell Donnelly, Mike Field, David Fowler, Gabriela Gomes, Brian Goodwin, Mike Impey, Gérard Iooss, Barbara Keyfitz, Greg King, Edgar Knobloch, Bill Langford, Alan MacKay, Jerry Marsden, Ian Melbourne, James Montaldi, Jim Murray, David Nelson, Terry Pratchett, Mark Roberts, David Schaeffer, Michelle Shatzman, Mary Silber, Jim Swift, Harry Swinney, and Randy Tagg. Their contributions range from fundamental research to classical scholarship, from pointing out ghastly errors in drafts to drawing computer graphics, from tracking down three-legged dogs to telling us about left-handed snails. Our debt to our predecessors is also obvious: it is explicit in the numerous quotes from their writings, and implicit in the many ideas we've stolen.

Fearful Symmetry isn't about technology, about gadgets, about gee-whiz breakthroughs that will revolutionise our lives. It's more of an exercise in what used to be called natural philosophy: a broad-brush picture of some of the more interesting of nature's marvels, drawn together from a unified viewpoint. Symmetry-breaking isn't the long-sought Theory of Everything (though physicists in search of such a theory make strong appeals to symmetry-breaking); but we believe it to be a Theory of Something, which may well be a better idea. Once you've become sensitized to that something, to the role that symmetry-breaking plays in nature's patterns, then you'll see the world around you with new eyes.

INS and MG
Coventry and Houston

1

Geometer God

One could perhaps describe the situation by saying that God is a mathematician of a very high order, and he used very advanced mathematics in constructing the universe.

Paul Dirac, *The Evolution of the Physicist's Picture of Nature*

'God Ever Geometrizes'

So (it is said) the Greek philospher Plato stated his belief that the physical universe is governed by laws that can be expressed in the language of mathematics. It isn't necessary to subscribe to the existence of a personal deity to observe that there are patterns and regularities in the world that we inhabit, and to wonder why; but Plato's statement is economical, pointed, and strikes at the heart of the matter. Nature's patterns are mathematical.

Numerous scientists throughout the ages have held similar beliefs. In 1956 Paul Dirac, one of the founders of quantum mechanics, visited the University of Moscow, and was asked to write an inscription on a blackboard, to be preserved for posterity. It was an honour reserved for only the very greatest visitors. Dirac knew this, and selected his personal credo: 'A physical law must possess mathematical beauty.'

Nearly all of today's science is founded on mathematics; indeed the state of maturity of a science is often judged by how mathematical it has become. Even biology, traditionally one of the least mathematical of the true sciences, has become much more so as it delves ever more deeply into the information-processing structures that surround the DNA molecule. Mathematics in its purest form – logic – lies at the heart of genetics, evolution, what it means to be alive, to be human.

Scientists use mathematics to build mental universes. They write down mathematical descriptions – models – that capture essential fragments of how they think the world behaves. Then they analyse their consequences. This is called 'theory'. They test their theories against observations: this is called 'experiment'. Depending on the result, they may modify the mathematical model and repeat the cycle until theory and experiment agree. Not that it's really that simple; but that's the general gist of it, the essence of the scientific method. The real thing, with its emotional commitment, priority disputes, and Nobel prize political infighting, also has sociological aspects. Those arise because science has to be performed by *people*, and people suffer from all sorts of distractions. The strangest aspect of this process is that it works. God *does* geometrize. Or, at least, humans often manage to convince themselves that She does. The growth of scientific understanding has been matched pace for pace by the development of mathematics; the two go hand in hand. The image of the Geometer God is powerful in art: for example, William Blake's painting *The Ancient of Days* of 1794 (Figure 1.1) shows the deity, dividers in hand, measuring up the universe for the act of creation.

The precursor to this book, *Does God Play Dice?*, examined *chaos*, the new mathematics of irregularity – the (sometimes only apparent) *absence* of pattern. *Fearful symmetry* is a 'prequel' rather than a sequel: it centres around the mathematics of regular pattern, which is conceptually simpler than chaos. Patterns in nature are a constant source of surprise and delight, and the dominant source of pattern in symmetry.

In everyday language, the words 'pattern' and 'symmetry' are used almost interchangeably, to indicate a *property* possessed by a regular arrangement of more-or-less identical units – for example the typographical pattern EEEEEEEEEEEE. Nature uses a similar pattern to design centipedes. Mathematicians use the word 'pattern' informally, and in much the same way; but they reserve 'symmetry' for a concept that is more precise than its everyday usage, and in some respects slightly different from it. The mathematician's view of symmetry applies to an idealized string of Es that is infinite both to left and right. It focusses upon the *transformations* that leave the entire string of symbols looking exactly the same. One such transformation is 'shift every symbol one place sideways'. Another is 'turn everything upside down'. The first transformation expresses the fact that each segment of a centipede looks the same as its neighbours; the second, that a centipede looks exactly the same in a mirror. There are other transformations that leave the entire string of symbols looking

Figure 1.1 The Geometer God plans His universe: William Blake's The Ancient of Days

exactly the same, such as 'shift two places sideways and turn upside down', but they are simply combinations of the two just described.

To a mathematician, an object possesses symmetry if it retains its form after some transformation. A circle, for example, looks the same after any rotation; so a mathematician says that a circle is symmetric, even though a circle is not really a pattern in the conventional sense – something made up from separate, identical bits. Indeed the mathematician generalizes, saying that *any* object that retains its form when rotated – such as a cylinder, a cone, or a pot thrown on a potter's wheel – has *circular symmetry*. One advantage of this approach is that it provides a quantitative approach to symmetry, allowing comparisons between different objects. For instance, an object that retains its form under only *some* rotations – such as a square, which loses its orientation unless it is rotated through a multiple of a right angle – can sensibly be described as being *less* symmetric than a circle.

One of the great themes of the past century's mathematics is the existence of deep links between geometry and symmetry. Blake's painting anticipates this in its powerful suggestion of bilateral symmetry. If the signature of a Dicing Deity is chaos, then the signature of a Geometer God is symmetry. Is it inscribed upon our world?

Indelibly. Our universe is an apparently inexhaustible source of symmetric patterns, from the innermost structure of the atom to the swirl of stars within a galaxy. Plato made mathematical regularity the keystone of his philosophy, and viewed reality as an imperfect image of an ideal world of pure forms. In Book VII of *The Republic* he offers a striking allegory of people imprisoned in a cave, able to see only shadows of the outside world on the wall:

> And do you see, I said, men passing along the wall, some apparently talking and others silent, carrying vessels, and statues and figures of animals made of wood and stone and various materials, which appear over the wall?
>
> You have shown me a strange image, and they are strange prisoners.
>
> Like ourselves, I replied; and they see only their own shadows, or the shadows of one another, which the fire throws on the opposite wall of the cave?
>
> True, he said; how could they see anything but the shadows if they were never allowed to move their heads?
>
> And of the objects which are being carried in like manner they would only see the shadows?
>
> Yes, he said.
>
> And if they were able to talk with one another, would they not suppose that they were naming what was actually before them?

Whether or not we embrace Plato's theory of forms – ideal shapes whose 'shadows' we see on the walls of our world, imagining them to be the primary reality – there are few of us who are unmoved by the extraordinary tendency of nature to produce mathematical patterns. How do such patterns arise? We shall concentrate on one very fundamental process of pattern-formation, known as *symmetry-breaking*. This is itself a paradoxical phenomenon: it occurs when a symmetric system starts to behave less symmetrically. In some manner – which we'll explore – symmetry gets lost. Curiously, the typical result of a loss of symmetry is pattern, in the sense of regular geometric form, because only seldom is *all* symmetry lost. An oddity of the human mind is that it perceives *too much* symmetry as a bland uniformity rather than as a striking pattern; although some symmetry is lost, pattern seems to be gained because of this psychological trick. We are intrigued by the pattern manifested in circular ripples on a pond (why circles?), but not by the even greater symmetry of the surface of the pond itself (it's pretty much the same everywhere). Mathematically, a uniform, featureless plane has a vast amount of symmetry; but nobody ever looks at a wall painted in a single colour and enthuses over its wonderful patterns. This perceptual quirk often makes nature's patterns seem more puzzling than they really are.

To those whose perceptions have been suitably sharpened, symmetry-breaking is everywhere. It's visible in the way people walk and horses trot, in the dew that glistens on spiderwebs at dawn, the ripples on a pond, the swell of ocean waves, the stripes of a tiger, the spots of a leopard. It evidences itself in less homely ways: the glittering facets of crystals, the ponderous vibrations of stars, the spiral arms of galaxies – perhaps even the gigantic voids and clusters of the universe, of whose presence science has only recently become aware.

Or so we shall argue.

Even the splash of a raindrop is symmetry-breaking in action. Let's begin with raindrops.

The Shape of a Splash

Our favourite oddball science book is *On Growth and Form* by d'Arcy Thompson. If you've never read this provocative and penetrating treatise, get a copy from somewhere – though be warned, part of its appeal is an outmoded charm, so don't take it too seriously. Thompson was a pioneer of the idea that there are mathematical features to biological form. Prominently displayed at the very front of his book there's a wonderful and slightly disturbing picture of a drop of milk hitting the surface of a bowl, filled with the same liquid; the splash is frozen by high-speed photography, for us to contemplate at leisure. When raindrops hit a puddle, or inkblots hit paper, they must do something similar. Have you ever wondered what shape a splash is?

It looks like a crown.

From the point of impact rises a smooth, circular ring, surprisingly thin-walled, curving gracefully outwards as it rises. But the ring doesn't remain circular: it breaks up into 24 pointed spikes (Figure 1.2). Why does it break up? Why 24? These are good questions. The spikes are (almost) regularly spaced. Why? That's another good question. For a time we'll accumulate questions; eventually we'll attempt to answer some of them. The spikes come to a sharp point; most have just thrown off a tiny rounded droplet of milk (why?), and the rest are about to. You can see traces of such spikes in the way water is flung out when raindrops hit a puddle. And you know inkblots are always spiky, that's how cartoonists draw them. If somebody showed you a circular inkblot, you'd never recognize it; you'd think it was just a black circle. Which is curious, remarkably

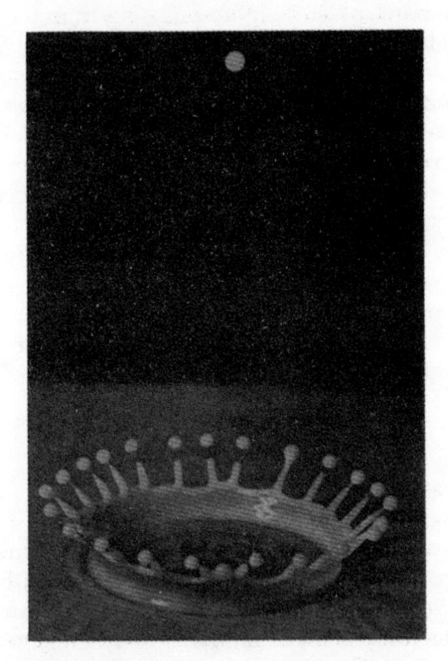

Figure 1.2 The symmetry of a splash

curious, because the drop of ink that produces the blot is (near enough) spherical, and the paper is flat. So what distinguishes the directions that lead to spikes from those that don't?

Focus your attention on the symmetry of the splash in d'Arcy Thompson's picture. It isn't perfect, but presumably that's due to slight imperfections in the shape of the original drop or the angle at which it fell. Maybe it was wobbling a little, maybe the milk in the bowl wasn't completely still. But the dominant feature, the spiky crown, doesn't look as though it's caused by such imperfections. You get the feeling that a *perfectly* spherical droplet would just give a *perfect* (and very probably also 24-pointed) crown! Let's assume this is true – for it is, in mathematical models of the process – and draw out the paradoxical consequences.

View the entire sequence of events from vertically above the bowl. A perfect droplet of milk has a perfect circular outline; and as it falls vertically downwards the outline remains circular. At all times, what you see has circular symmetry. If you now change viewpoint, and imagine your eye positioned at the centre of that circle, looking horizontally, then you won't be able to tell in which direction your eye is pointing. The droplet looks identical in any horizontal direction. So does the bowl, at least if we use a circular bowl and drop the milk into the middle of it. Thus we have a *cause*, the falling droplet; and the entire system of droplet, milk, and bowl has perfect circular symmetry – it looks the same in all horizontal directions.

What about the *effect* – the splash?

That *doesn't* have circular symmetry. It looks different, depending on which direction you view it from. Imagine placing yourself so that its nearest part is a spike; now walk round a few degrees, and the nearest part is a gap instead. The two views are similar, but not identical: where one has spikes, the other has gaps.

Where did the symmetry go?

Curie's Principle

The great physicist Pierre Curie is best remembered for his work, with his wife Marie, on radioactivity, leading to the discovery of the elements radium and polonium. But Curie is also remembered for his realization that many physical processes are governed by principles of symmetry. In 1894, in the *Journal de Physique Théorique et Appliquée,* Curie gave two logically equivalent statements of a general principle from the folklore of mathematical physics:

- If certain causes produce certain effects, then the symmetries of the causes reappear in the effects produced.
- If certain effects reveal a certain asymmetry, then this asymmetry will be reflected in the causes that give rise to them.

Curie's Principle (we'll refer to it in the singular, since its two statements are equivalent) is more subtle than it seems: like the utterances of politicians, its truth depends on how you interpret it. Let's begin with the simplest interpretation, which we can paraphrase as 'symmetric causes produce equally symmetric effects'.

At first sight, the principle is 'obviously' true. If a planet in the shape of a perfect sphere acquires an ocean, then we expect that ocean to be a perfect sphere as well, so it should coat the planet to the same depth everywhere. If the planet rotates, then we expect the ocean to bulge at the equator, but to retain circular symmetry about the axis of rotation. The flow of air past a symmetric obstacle will be symmetric. A rubber cube compressed by equal forces perpendicular to each face will obviously remain a cube, albeit a smaller one. The flow of fluid in an apparatus with circular symmetry will have circular symmetry. Isn't that right? After all, what else could happen?

Let's examine one example in more detail: the flow of air past an obstacle – in this case, a jumbo-jet. In the Kensington Science Museum in London there's an engineering model of a jet airliner, which during its design was used in a wind-tunnel to study the flow of air around the craft. An aircraft is bilaterally symmetric; that is, its left and right halves are mirror-images of each other. The engineers therefore only built half the model: they made a left-hand half of the plane and mounted it against a flat wall. The assumption behind this is that the flow of air past the aircraft must be bilaterally symmetric as well.

Here's a second example, also from industry. An American oil company was modelling the flow of oil in a system of production wells. It set up a mathematical model – a system of equations designed to mimic the main features of the real system – in which a hexagonal honeycomb of production wells extracted the oil, and injection wells at the centre of each hexagon pumped water in to displace it. As is common in industrial problems, a computer was used to set up and solve the mathematical model. In order to save computing time, the company assumed that the flow-patterns of the oil and water would have the same hexagonal symmetry as the wells; that let them restrict their analysis to a triangular region forming one twelfth of a hexagon (Figure 1.3).

Are the symmetry assumptions justified in these two cases?

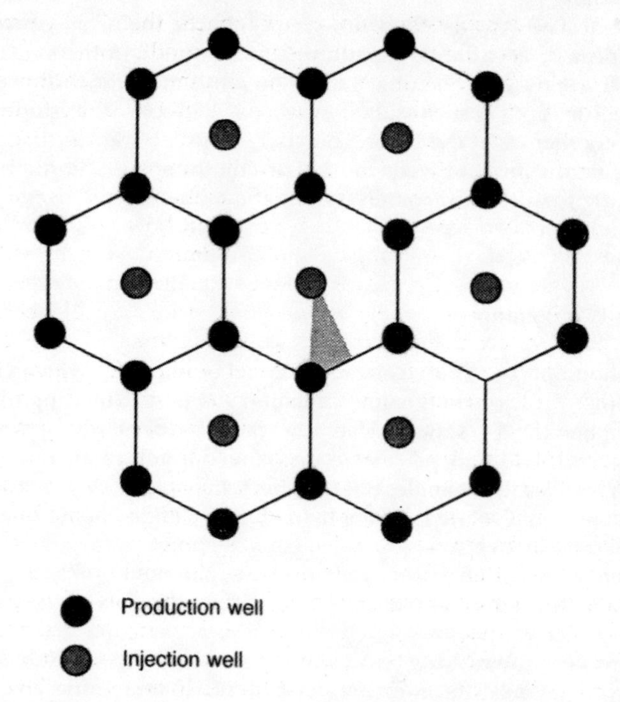

Figure 1.3 Hexagonally symmetric model of an oilfield. The flow of oil in the shaded triangle determines the rest of the flow 'by symmetry'. Or does it?

Let's think about the airliner. Imagine a complete airliner, both halves present, perfectly symmetric. Air flows past the craft according to the laws of aerodynamics. It's a mathematical fact, inherent in the form of those laws, that they are invariant under reflection. That is, if you reflect the aircraft, and simultaneously reflect the flow past it, then the result again obeys the laws. Now let's use the fact that the aircraft is symmetric under such a reflection: it means that after the reflection, the aircraft looks exactly the same as it did before. So the flow past it must look the same as before, too. But the flow past it has been reflected! We deduce that the flow must also be invariant under reflection.

A similar argument applies to the hexagonal array of oil wells. Hexagonal symmetry means that the array looks the same after various transformations: rotations through multiples of 60°, reflections in symmetry lines, translations sideways from one hexagon to the next. The laws of fluid flow imply that if the array of wells is transformed, and the flow-pattern is transformed in the same way, then it yields a physically valid flow. But the array of wells is symmetric under such transformations: it looks the same after applying them as it did before. So after each such transformation the flow past the array of wells must also look the same as it did before, implying that it's hexagonally symmetric too.

Convinced?

Curie Was Wrong

You shouldn't be. You've already seen an example for which Curie's principle – at least, in this interpretation – is false. The drop of milk. Remember the 24 spikes? Here the cause is circularly symmetric: drop, bowl, and milk all remain unchanged if you rotate everything through an arbitrary angle. But the effect, the splash, does *not* remain the same if you rotate it through an arbitrary angle. Some rotations, for example, move spikes to gaps, gaps to spikes. The only rotations that leave the crown unchanged are those that move spikes to spikes: rotations through multiples of $\frac{360}{24} = 15°$. The effect has less symmetry than the cause.

However, down at the bottom of the splash, when it first begins to rise from the milk, the effect *does* look circularly symmetric. so Curie's Principle seems to hold at first, but subsequently it goes wrong, as the initially circular symmetry breaks to 24-fold rotational symmetry. The splash captured by the high-speed camera isn't just an example of broken symmetry: it's a frozen representation of the entire *process* of symmetry-breaking. That's one reason why it's so intriguing.

Along with Curie, we're in trouble. We have a logical argument that seems to prove his principle is right, and an example that seems to prove it's wrong. Not such an unusual situation in scientific research! Either the example is wrong, or the argument is – maybe both, of course. Our job is to find out which. Here we can apply a direct approach: put the example up against the argument, and see which wins.

Here we go. The cause (droplet) has circular symmetry. That is, if we rotate everything through an arbitrary angle, nothing changes. Consider a splash (effect). Its form is derived from that of the droplet by applying the laws of fluid dynamics. Those laws are invariant under rotations; that is, if we rotate the drop through an arbitrary angle and rotate the splash through the same angle, we get a physically possible sequence of events. Seems watertight so far.

Now: circular symmetry implies that the rotated drop looks just like the original. Therefore the rotated splash looks just like the original.

But – *it doesn't*.

Hmm. There's a glitch in the logic, somewhere.

Let's think about it this way. Suppose we use computer graphics to make another photograph, but taken from a slightly different angle: to be precise, rotated through $7\frac{1}{2}°$, so that the spikes on the crown move to where the the gaps were in the original. If what we've been arguing so far is right, then this second picture represents a physical impossibility.

If you're happy with that, here's a question. How do you know that Figure 1.2 is the original picture from d'Arcy Thompson, as we've been assuming? Might it not be the fake, physically impossible computer picture? How can you tell? The only difference between the two is that what's North in one is $7\frac{1}{2}°$ East of North in the other.

But nobody's *marked* North ... and in a circularly symmetric system there's no reason to choose any particular direction to *be* North.

This is puzzling. A slightly rotated splash, with its spikes where once there were gaps, seems to be just as valid an effect as the splash that actually occurs. Can the droplet have *more than one effect*?

In the real world, no: something definite has to happen. You don't get two splashes at the same time. But in the mathematics, yes. Both splashes, the original and its rotation, are valid solutions to the same equations; valid consequences of the same physical laws. Instead of a single effect, we have a whole *set* of possible effects: all the different rotations of the 24-pointed crown. Our argument from symmetry doesn't prove that 'the' effect is symmetric under rotation: what it proves is that if *an* effect is rotated, then the result is a physically possible effect – but it could be a *different* one. The logical fallacy is the assumption that each cause produces a unique effect; the argument breaks down if several distinct effects are equally possible.

It may sound unlikely that several possible effects can arise from a single cause, but symmetric systems are like that. Symmetry means that, given some possible effect, all symmetrically related effects are also physically possible. Take the aircraft, for example. What the argument really shows is that, given any possible flow of air past the craft, its mirror image is also a possible flow (Figure 1.4). *They don't have to be the same.*

One of our acquaintances, an aircraft engineer, tells us that if you sit right at the back of one popular make of commercial airliner, then you'll notice that every five minutes or so it seems to twitch sideways a couple of feet. Five minutes later it twitches the other way. This is

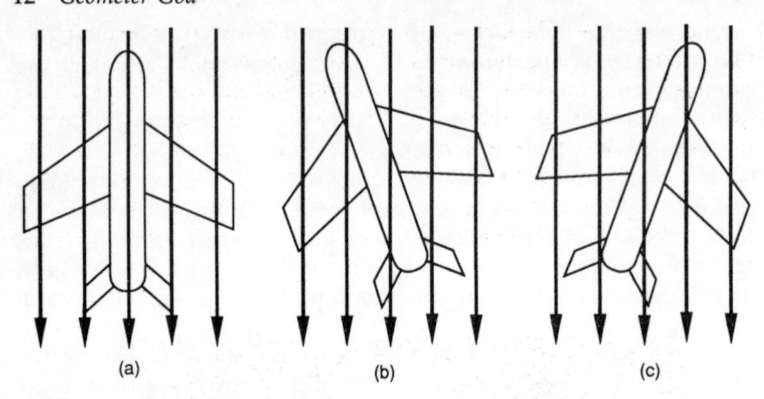

Figure 1.4 Flow past a bilaterally symmetric aircraft. (a) need not happen in practice; but if (b) does, so must (c)

because the flow of air past the aircraft is *not* bilaterally symmetric. The plane flies very slightly crabwise, its tail a foot or so to one side of the line of flight. Any change – a bit of turbulence, the automatic pilot adjusting the control surfaces – and it may twitch across to the symmetrically related position. The engineers who designed the aircraft know all about this effect. It's perfectly safe, and so tiny that it has no practical implications. You have to know what to look for before you can spot it happening. As a practical matter, *both* flows are extremely similar to each other, and to the perfect symmetric flow that the wind-tunnel experiment, by using half an aircraft, forces to occur. But it's an interesting case where our 'Curie Principle' intuition about symmetry is violated, not only in theory, but in practice.

A final question which may be bothering you. It all seems a complicated way to design an aircraft. Why doesn't it just fly straight ahead?

It can't.

Loss of Stability

To see why an airliner may not be able to fly straight, we must explain what causes symmetry to break.

A simple, more familiar example is helpful at this point. Think of a perfect circular cylinder, say a tubular metal strut, being compressed by a force. What happens? Nothing much at first, but if the force becomes sufficiently large, then the strut will buckle. The buckling is *not* a consequence of lack of symmetry caused by the force: even if the force is directed along the axis of the tube, preserving the rotational symmetry about that axis, the tube will still buckle. Buckled cylinders cease to be cylindrical – that's what 'buckle' means. Figure 1.5 shows the result of an experiment in which a metal cylinder, sandwiched inside a slightly larger glass one to prevent it buckling too far, is compressed from its ends by a uniform force. An elegant, symmetric pattern of identical dents appears, in a fairly random order.

Figure 1.5 Buckling of a cylinder into a symmetric pattern of dimples, under pressure from the ends. The cylinder is set inside a slightly larger glass one to prevent large buckles

Mathematically, even when it is compressed by very large forces, a cylindrical strut can remain cylindrical. There does exist a solution to the mathematical model in which there is no buckling. But mathematically, you can balance a chain on end, provided each link sits precisely above the one below it. The Indian rope trick notwithstanding, this is impossible in the real world. So mathematical *existence* of solutions to equations isn't enough.

The missing ingredient is *stability*. Natural systems must be stable; that is, they must retain their form even if they're disturbed. A pin lying on its side is stable and can occur in the real world. A pin balanced on end is *theoretically* possible, but it does not occur in the real world, because it is unstable: the slightest breath of wind and it topples. The same is true of a vertical chain, and of an unbuckled strut under enormous compression. What actually happens when a strut buckles is that when the compression reaches a critical load, the unbuckled state becomes unstable. The strut then seeks a nearby stable position; and that's buckling.

Here's an experiment for you to try, either with real apparatus or just in your mind. Take a plastic ruler, and hold it flat between your hands: then compress the ends. The ruler has top/bottom symmetry (provided it's flat), and in its fully symmetric state it's horizontal. If you don't press too hard, that's what happens. But greater compression makes the ruler buckle – say upwards. By symmetry there ought also to be a position where it buckles downwards; and there is. If you've got three hands, you'll find that in between the two buckled states there's a position where the ruler *tries* to be horizontal, in the sense that it takes very little pressure to keep it there; but it keeps flipping away when you release it. That's the fully symmetric state again, but now it's unstable. As the compressive forces increase, the symmetric state loses stability, and the ruler has to buckle to an asymmetric position.

The same is true of the crabwise-flying aircraft: straight-ahead flight is unstable. If it tilts an inch away from straight, it tends to move further away, rather than returning, thanks to the very complex interplay of forces upon its airframe. As it happens, there's a stable position about a foot to either side of the centre, thoughtfully provided by the design engineers.

We can also see why the milk splash starts out having circular symmetry, but then loses it, forming spikes. We've already observed that initially the growing ring of milk has a circular form, and that the spikes appear higher up. Presumably the ring becomes unstable when it grows too high, and it – buckles. Just like the sphere and cylinder, it goes wavy. The detailed fluid dynamics must select the form with 24 waves, though we can't confirm that without doing a very nasty calculation. However, by using the general mathematics of symmetry-breaking, we *can* predict that the remaining symmetry will be that of a regular polygon, although the number of sides – here 24 – can't be deduced from symmetry-breaking alone. In short, while the detailed structure of the crown is a surprise, its general features, especially its symmetries, are not.

Curie was right in asserting that symmetric systems have symmetric states – but he failed to address their stability. If a symmetric state becomes unstable, the system will do something else – and that something else need not be equally symmetric.

Curie Was Right

We shouldn't be too pleased with ourselves, having caught a Nobel-winning physicist making a mistake: we may just have misunderstood what he was trying to tell us. There's a disturbing gap in the story at this point. We've said that *mathematically* the laws that apply to symmetric systems can sometimes predict not just a single effect, but a whole set of symmetrically related effects. However, Mother Nature has to *choose* which of those effects she wants to implement. How does she choose?

The answer seems to be: imperfections. Nature is never *perfectly* symmetric. Nature's circles always have tiny dents and bumps. There are always tiny fluctuations, such as the thermal vibration of molecules. These tiny imperfections load Nature's dice in favour of one or other of the set of possible effects that the mathematics of perfect symmetry considers to be equally possible.

Take a perfect sphere, and compress it with a uniform radial force. That's a fancy way to say 'try to squash a ping-pong ball'. When the sphere buckles, it develops a dent somewhere. According to the mathematics, the dent is circular in shape, and indeed the buckled sphere retains circular symmetry about some axis (Figure 1.6). Don't expect to see anything as perfect as this if you squash a real ping-pong ball: the picture is an ideal case. In principle, if a buckling pattern can occur, than any rotation of that pattern can also occur. That is, the axis of rotational symmetry, and the dent at one end of it, can be in any place you like. The *shape* of the buckled sphere is the same in each case, but its *position* is not. Not only is Figure 1.6 a possible form for a buckled sphere: so is anything you can get by rotating it. A real sphere, however, is never exactly spherical. It may, for example, be slightly thinner at one place than at another. The dent

Figure 1.6 When a sphere first buckles, it has circular symmetry

is more likely to occur at such a weak spot. That's how Mother Nature loads the dice.

In other words, the issue is the relation between a mathematical model and the reality that it is supposed to represent. Nature behaves in ways that look mathematical, but nature is not the same as mathematics. Every mathematical model makes simplifying assumptions; its conclusions are only as valid as those assumptions. The assumption of perfect symmetry is excellent as a technique for deducing the conditions under which symmetry-breaking is going to occur, the general form of the result, and the range of possible behaviour. To deduce exactly *which* effect is selected from this range in a practical situation, we have to know which imperfections are present.

In this sense, Curie was absolutely right. If we see a ping-pong ball with a dent in one side, we are right to deduce that something happened to it that was not spherically symmetric; that the asymmetric effect we observe must have had an asymmetric cause. However, that asymmetry might be just a tiny fluctuation in an otherwise perfectly symmetric setting. So whether Curie's Principle is true depends on what questions we ask.

History is littered with examples where scientists and philosophers have misapplied Curie's Principle, seeking *large-scale* asymmetries in causes, to account for large-scale asymmetries – such as patterns – in effects. For example, until very recently astronomers thought that the spiral arms of galaxies were caused by magnetic fields. As we'll see in chapter 6, they're beginning to think that the spirals are the result of gravitational symmetry-breaking. But Curie's Principle doesn't say that the size of the asymmetry is comparable in the cause and the effect. As we shall see, when systems have symmetry, there's a good chance that the symmetry may break. When it does, very tiny asymmetries play a crucial role in selecting the actual outcome from a range of potential outcomes.

Trucks and Trees

Once you've become sensitized to the possibility of symmetry-breaking, you see it everywhere. A few summers ago one of us (INS) was driving along a freeway in up-state New York. Ahead was a large truck, with two mud-flaps at the rear. They were flapping - as all good mud-flaps should. But they weren't flapping in unison. When the left-hand flap was moving forwards, the right-hand one was moving backwards, and *vice versa*. An engineer would note that the oscillations were 180° out of phase. A physicist would observe that the oscillations were caused by vortex-shedding: the truck was leaving a train of tiny tornados in its wake, peeling off in turn to the left and the right, and wiggling the flaps as they passed. But what your humble author saw was an example of symmetry-breaking. The truck and its arrangement of flaps was, near enough, left-right symmetric; but the motion was asymmetric: the left-hand flap and the right-hand flap were not performing identical motions.

In fact the pattern of vortices has its own symmetry, but of a different kind from that of the truck that produces it. The truck is symmetric under a reflection that interchanges left and right (Figure 1.7a); the vortex train that it sheds is symmetric under a *glide reflection*, which interchanges left and right but also moves a suitable distance in the direction of the truck's motion (Figure 1.7b).

Here's another broken symmetry, witnessed by the same author on a previous trip. In northern California grow huge trees, redwoods and sequoias. The trunk of a tree is approximately cylindrical, and we

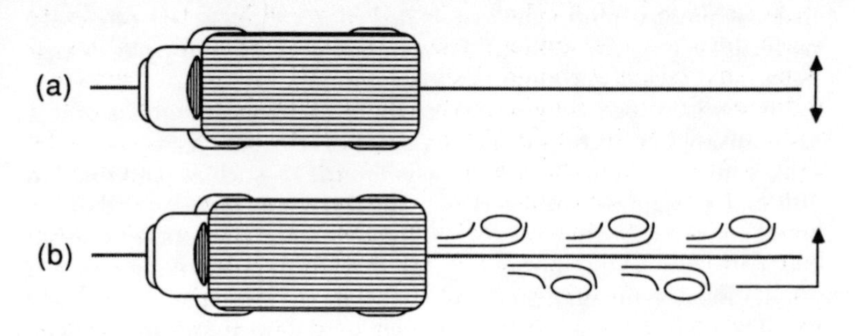

Figure 1.7 (a) Bilateral symmetry of a truck. (b) Glide-reflection symmetry of the vortex train that it produces

may assume that it has exact cylindrical symmetry. The symmetries of a cylinder are of three kinds: rotation, translations, and reflections. If you rotate a cylinder about its axis (Figure 1.8a) it looks exactly the same as before; and the same is true if you translate it in the direction of its axis (Figure 1.8b). To be precise, translational symmetry holds only for an infinitely long cylinder; but it is valid to a good approximation for a sufficiently long one. There are also two distinct types of reflectional symmetry (Figures 1.8c and 1.8d): either in a vertical mirror or in a horizontal one.

It seems plausible that the pattern of bark on the tree should have similar symmetry to the tree itself. Now a pattern of bark with (a good

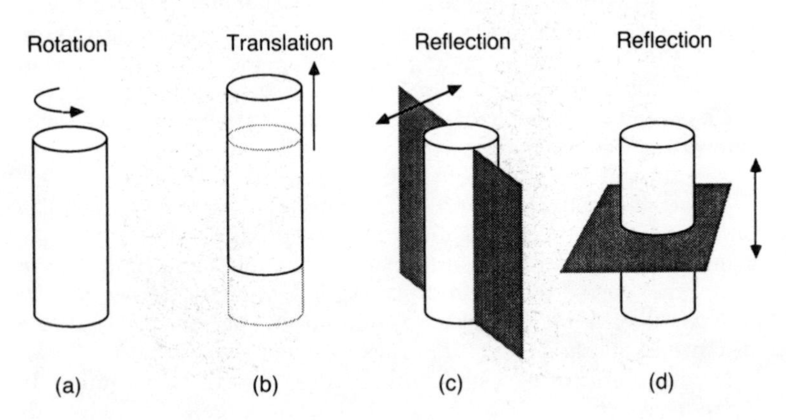

Figure 1.8 Symmetries of a cylinder

approximation to) full cylindrical symmetry will have to look pretty much the same after any of those rotations, translations, and reflections. That means it should resemble Figure 1.9a, with the grooves in the bark running roughly vertically – as they do on most trees.

But on *some* of the Californian trees, you'll see a *spiral* pattern in the bark (Figure 1.9b). The spiral still has some symmetry, but of a different kind. If you rotate a helical spiral, *and* translate it, then it looks the same. So the symmetry of a spiral is a mixture of rotation and translation, known as a *screw*. Indeed, this is the reason why a carpenter's woodscrew works: as it rotates *and* goes deeper into the wood (or translates) it fits into the same hole. Real screws are tapered so they enlarge the hole slightly as they go, for a tight fit; but a bolt

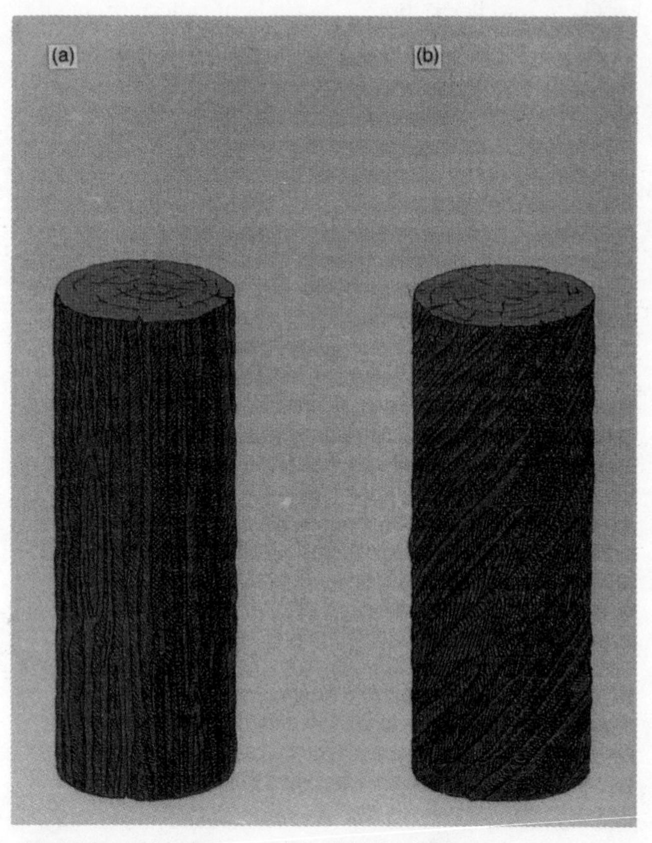

Figure 1.9 (a) Straight bark on a sequoia. (b) Spiral bark

with a helical thread has exact symmetry of this kind. Did something strange happen to those trees that developed spiral bark? Pesticides, a bad winter, drought? Or should we *expect* spiral patterns as well as perfectly symmetric ones? From our new-found viewpoint, there's nothing surprising about trees with spiral patterns to their bark. If the perfectly symmetric pattern represents an unstable development, then tiny disturbances will cause the symmetry to break. Spirals are one of the common ways to break cylindrical symmetry, so a spiral pattern could very well develop instead.

Spiderwebs

If you get up early on a spring morning, and go outside, everything sparkles like a jeweller's shop. The sunlight reflects from countless droplets of dew. Spiders' webs are especially attractive – and only on a dewy morning do you realize just how many spiders there are in the world, for the webs are everywhere.

Why do you see individual droplets on the threads of a web? Silly question! You see droplets everywhere else, why not on webs? A drop of dew falls, hits, sticks; then another does, and so on. That's all there is to it, isn't it? Not entirely. Look closely at a dew-bedecked web, and you'll see something interesting: the droplets are (pretty much) evenly spaced along the threads. That's especially true along the long straight threads that tie the edges of the web to bushes and walls.

Why are they equally spaced? Here's one explanation, the one that appeals to symmetry-breaking. Imagine an idealized thread of spider silk, an infinite, straight line. Imagine that dew falls on to the thread and spreads, coating it. The symmetries of the thread are all rigid motions of the line; that is, all translations and all reflections. Moreover, it has a circular cross-section. If symmetry isn't broken, then the coating of dew must have the same symmetries as the thread; which means that the entire thread will be coated evenly in water, and the dew will create a wet cylinder, of uniform cross-section, with the thread running along its centre. Why doesn't that happen? The fully symmetric state is unstable, because the surface tension of the water is compressing it along the direction of the thread, like a very tall stack of books tied with elastic. So the water bulges here and there. However, it generally seems to be true that when symmetry breaks, quite a bit of it tends to be retained. Both the buckled sphere and the buckled cylinder have a lot of pattern left, for example. So some translations and some reflections may not be broken. If a translational symmetry remains, then the entire pattern of dewdrops must repeat at regular intervals; if there's a reflectional symmetry, then those component patterns will be bilaterally symmetric. And that's what we find: regularly spaced blobs, nearly spherical and each blob left-right symmetric.

We could spend some considerable time describing the physics of this situation, and fleshing out the abstract symmetry-breaking 'explanation'. However, we just want you to look at dew-bedecked spiderwebs with fresh eyes, to see them as yet another example of pattern formation caused by broken symmetry. So we'll move on.

Honeycomb Lakes

In some parts of the world there are curious flat mounds of rocks – geologists call them stone nests – arranged in a roughly hexagonal pattern like a honeycomb (Figure 1.10). Why? Originally the stones were on the bed of a large, shallow lake. The Sun's rays heated the lake, giving rise to currents in the water. Now a large lake is approximately symmetric under translations in any direction, as well as rotations. If no symmetry were broken, the flow of water would also be symmetric under all translations and rotations – which means no flow at all! All that would happen is that the heat would be *conducted* through a stationary lake.

To find out what really happens, take a frying-pan filled with a shallow layer of water (the lake) and sit it on the stove (Sun). The

Figure 1.10 Hexagonal pattern of stone nests, caused by convection

depth of water should be no more than half a centimetre or so. Turn on the stove and heat the pan *very* gently. (Be careful, and if you are under the age of 21 or cohabiting with another human being ask permission first. A cast iron pan is best, to distribute the heat uniformly across the base of the pan. It may be wiser to treat this as a thought experiment: at any rate, don't expect too spectacular a result – proper laboratory equipment works better.) What happens?

You get strange cellular patterns.

The physical reason is that hot fluid tends to rise. As the temperature increases, the hot layer on the bottom is trapped beneath a layer of colder, denser fluid. This situation is unstable, and is destroyed by the onset of convection. Hot fluid rises in some regions, cold fluid descends in others. A cellular pattern of moving fluid, known as Bénard cells after the French scientist who discovered them, appears. Sometimes the pattern consists of parallel rolls, sometimes it is a honeycomb array of hexagons. In a real pan the symmetry is approximate and the pattern is rather irregular, but in an idealized infinite pan you get perfect honeycombs (Figure 1.11). These patterns still have a great deal of symmetry, but less than that of the pan.

Stone nests are relics of a similar process occurring in the lake. Over long periods of time, convection cells in the water clump the stones together: the approximate hexagonal pattern of the nests reflects that of the convection cells.

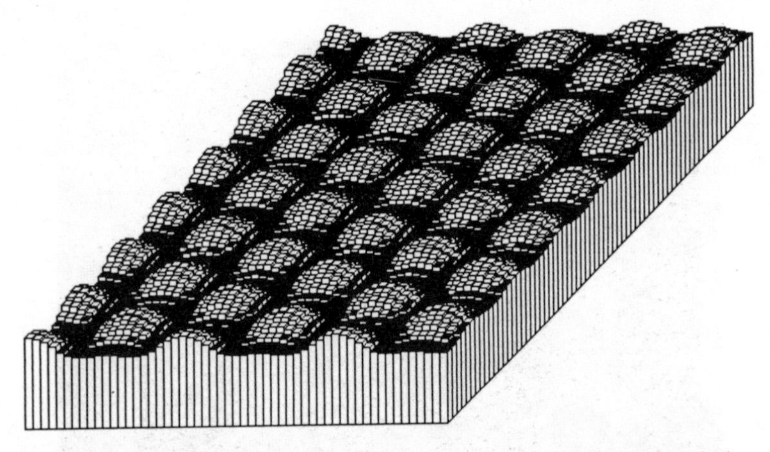

Figure 1.11 Computer-generated honeycomb pattern in a mathematical model of convection. The height of the surface represents the vertical speed with which fluid is moving

The truth is thus more complex than Curie's Principle suggests. A symmetric system will take up an equally symmetric state, *except when it doesn't!* It might seem that this just says that Curie was right except when he was wrong, but by analysing the conditions under which symmetry-breaking occurs we can give the idea some genuine content. To do that, we need to make the idea of symmetry precise, and that we do in the next chapter.

Corn Circles

There's a moral to the tale, and here's as good a place as any to make it. Our discussion of Californian tree bark doesn't *explain* how the spiral develops: we still don't know if it's caused by climate, genetics, a virus, or whatever. But we've understood an important point: spirals are a likely development, whatever the cause. If you want to find the cause, don't get too obsessed with spirals!

California is a place of extreme beauty, but also a place of weird cults and lifestyles. It's not the only place with that problem. As we write, the British media are awash with bizarre tales of 'corn circles'. These are remarkable depressions that occur, apparently spontaneously, in fields of standing corn (Figure 1.12). The simplest version is perfectly circular; other more elaborate shapes include concentric rings, circles with four smaller circles arranged around them, equally spaced, or entire chains of circles. The circles generally appear on still nights in large, flat fields.

Figure 1.12 Hoax, UFO, or broken symmetry? The enigma of corn circles

There are lots of theories. One is that they're the imprints of UFOs – after all, UFOs are circular, aren't they? Nobody has yet explained what benefit aliens get from landing in a cornfield, though. Another is that they're the result of vortices in still air – like smoke-rings. Smoke-rings are naturally circular. A third is that they're the result of electrostatic charges building up in the heads of corn, causing it to collapse like a tall stack of dominoes when you press too hard on the top. A fourth is that they're hoaxes, the lads from the local Young Farmers' Club out on a drunken spree: all you need is a peg and a rope, and you can soon trample a circle.

The interesting feature of almost all discussion of these phenomena is that they always focus on the circular shape. How could natural causes produce something as regular as a circle? It's always seen as the key to the enigma.

If you know about symmetry-breaking, however, you soon realize that the circularity is the one thing that *doesn't* need explanation! Think of it this way. A large, flat field is a very symmetric thing: it's not far removed from that mathematical ideal, an infinite plane. If the boundaries of the field are far away, the local geometry doesn't alter if you translate the field sideways a bit, reflect it in some line, or *rotate* (aha!) it. This is true even more of the corn: modern agricultural techniques lead to considerable uniformity in height, strength, and yield.

Suppose something happens to cause this symmetry to break. If the source of asymmetry is something that occurs at a point – for example, it may be nothing more than a bare patch of rocky ground – then the translational symmetry will be broken; but the rotational symmetry will remain. There need not even *be* a source at all: if the uniform state becomes unstable, symmetry-breaking will occur, and rotational symmetry is a strong possibility. What's the simplest geometric form with rotational symmetry?

A circle.

In symmetry terms, the process is completely analogous to the creation of circular ripples on a pond, by throwing in a stone. The pond has the symmetry of an infinite plane; the stone breaks symmetry at a point. The physical mechanisms are probably quite different, of course; and (so far) nobody has made a fortune writing books claiming that ripples on ponds are evidence of alien visitations.

The explanation of corn circles via electrostatic forces takes off from this point. The explanation by vortices applies similar reasoning, not to the field of corn, but to the atmosphere above it, although its devotees don't express it in terms of broken symmetry. The sharp edges of the circles are also no mystery: it takes a very specific amount of force to break a stalk of corn, and with modern farming techniques, stalks of corn are almost identical, so it takes the identical force to break them all. The boundary between regions where the force (be it of atmospheric, electrostatic, or extraterrestrial origin) is large enough to break the corn, or not, is of necessity sharp. Indeed when storms lay waste to cornfields, they tend to flatten it in well-defined, though irregular, areas. Having said all this, without doubt some circles are due to the efforts of the Young Farmers' Club, especially now that the phenomenon is attracting such coverage. An explanation of why aliens find it so amusing to land in cornfields remains elusive – maybe ET is trying to phone home with a very large circular satellite dish.

What have we learned by applying the ideas of symmetry-breaking? That the circular form is inherent in the flat, uniform nature of the surrounding field. That the same psychological trick is twisting our perceptions: we see the symmetry of the circle, we ignore the even greater – but blander – symmetry of the field of corn, and wonder where the circles come from. Corn circles are probably no more enigmatic than pond circles. The difference is that we can't see the stone being flung in, and we haven't yet worked out the details of the physics. Even so, we've learned something very valuable indeed: *not to be unduly impressed by the fact that corn circles are circular.*

If only the news media would do the same.

2

What Is Symmetry?

> Some years ago I worked out the structure of this group of operators in connection with Dirac's theory of the electron. I afterwards learned that a great deal of what I had written was to be found in a treatise on Kummer's quartic surface. There happens to be a model of Kummer's quartic surface in my lectureroom, at which I had sometimes glanced with curiosity, wondering what it was all about. The last thing that entered my head was that I had written a paper on its structure.
>
> Sir Arthur Stanley Eddington, *The Theory of Groups*

Symmetry appeals to artist and scientist alike; it is intimately associated with an innate human appreciation of pattern. Symmetry is bound up in many of the deepest patterns of Nature, and nowadays it is fundamental to our scientific understanding of the universe. Conservation principles, such as those for energy or momentum, express a symmetry that (we believe) is possessed by the entire space–time continuum: the laws of physics are the same everywhere. The quantum mechanics of fundamental particles, a crazy world in which a proton can be 'rotated' into a neutron, and whose laws must reflect this possibility, is couched in the mathematical language of symmetries. The symmetries of crystals not only classify their shapes, but determine many of their properties. Many natural forms, from starfish to raindrops, from viruses to galaxies, have striking symmetries.

But what, exactly, *is* symmetry? Let's consult an expert. Many of the greatest mathematicians have produced 'popular' works, to present their subject to a broad audience. Felix Klein wrote several popular books, including *Famous Problems of Elementary Geometry*, based upon his Easter lectures of 1894. Henri Poincaré wrote two bestsellers on the philosophy and methodology of science – *Science and Method* and *Science and Hypothesis*, books that remain in print to this day. David Hilbert, at the peak of his career the leading mathematician in the world, gave a radio broadcast to mark the International Congress of Mathematicians held in Paris in 1900. Hilbert always maintained that you don't really understand something until you can explain it to the first person you meet in the street. That's probably taking things a bit *too* far, but it makes it clear where Hilbert's sympathies lay.

In 1952 one of Hilbert's students, a distinguished mathematician named Hermann Weyl, was about to retire from the Institute for Advanced Study at Princeton. Weyl continued his mentor's tradition, and gave a series of public lectures on mathematics. His topic, and the title of the book that grew from his talks, was *Symmetry*. It remains one of the classic popularizations of the subject, and no book that follows in its footsteps can ignore it. Some of Weyl's greatest achievements had been in the deep mathematical setting that underlies the study of symmetry, and his lectures were strongly influenced by his mathematical tastes; but Weyl talked with authority about art and philosophy as well as mathematics and science. Here's how he began his first lecture:

> If I am not mistaken, the word *symmetry* is used in our everyday language with two meanings. In the one sense symmetric means something like well-proportioned, well-balanced, and symmetry denotes that sort of concordance of several parts by which they integrate into a whole. Beauty is bound up with symmetry.... The image of the balance provides a natural link to the second sense in which the word symmetry is used in modern times: *bilateral symmetry*, the symmetry of left and right, which is so conspicuous in the structure of the higher animals, especially the human body. Now this bilateral symmetry is strictly geometric and, in contrast to the vague notion of symmetry discussed before, an absolutely precise concept.

It's symmetry in the second, mathematical, sense, that interested Weyl in his professional work – although, as an intellectually sensitive person, he was by no means immune to the charms of the first sense. And it is symmetry in the mathematical sense that we must come to terms with now, even though our motives for being interested in it may well be mainly aesthetic. Beauty is a driving force for many mathematicians, but they like to delve below its surface and make it precise, even though the results of their excavations may not possess the same direct appeal to the rest of the human race.

It took humanity roughly two and a half thousand years to attain a precise formulation of the concept of symmetry, counting from the time when the Greek geometers made the first serious mathematical discoveries about that concept, notably the proof that there exist exactly five regular solids. Only *after* that lengthy period of gestation was the concept of symmetry something that scientists and mathematicians could *use* rather than just admire.

Changed, Yet the Same

In everyday usage, symmetry is an aesthetic property; but only a foolhardy mathematician would attempt to formalize such a viewpoint. Is there a mathematical concept which captures at least some important features of symmetry in the conventional sense, but which is amenable to rigorous analysis? There is. We've already prepared the ground, with our study of a splash of milk, so it's much easier for us to see what eluded mathematicians for so long. For their purposes, a symmetry isn't a *thing*; it's a *transformation*. Not any old transformation, though: a symmetry of an object is a transformation that leaves it *apparently* unchanged. We say 'apparently' because, although after the transformation the overall form of the object is the same as it was before, the object itself has moved. It must have done, otherwise the transformation wouldn't be interesting!

For example, consider an 'idealized' human figure, whose left side is exactly the same as the right side. Exactly? Well, if your left foot looked *exactly* the same as your right foot, you'd need two right slippers and it wouldn't mater which foot you put which slipper on. No, the two sides are not exactly the same: you have to flip one over to get the other. This is why the reflection of a human being in a mirror looks like a human being. In real humans there are subtle differences – for example, the heart of a mirror-person is on the wrong side, and a world populated by mirror-people would be predominantly left-handed. We'll leave the chemistry and biology of animal asymmetry for chapter 7, where it fits naturally into the development of our theme. Our immediate objective is to formulate a mathematical model that captures the perceived bilateral symmetry of the human form.

To recap, when we say that 'the left side is the same as the right', we mean that if you reflect the left half in a mirror, you obtain the right half. Now, reflection is a mathematical concept – but it isn't a geometrical shape, or a number, or a formula. It's a transformation, a rule for moving things around. Imagine an artist's drawing of the human body, with perfect bilateral symmetry (Figure 2.1). Draw a line straight down the middle, and flip the figure over so that this line

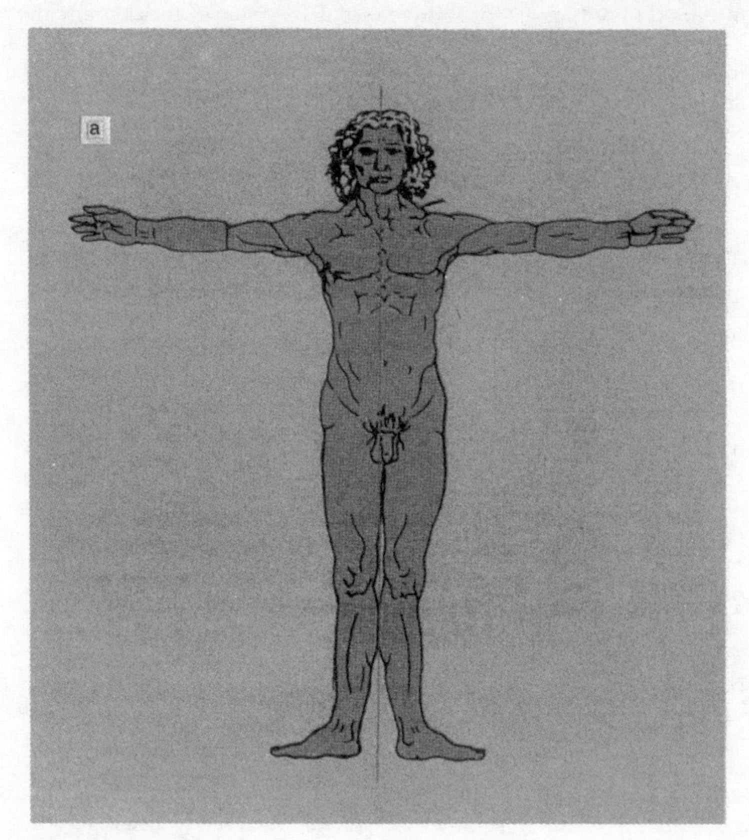

Figure 2.1 Symmetry axis of the human body

doesn't move, but the left and right sides are interchanged. Then – even though individual points have moved to different positions – the shape looks the same as it did before. We call the line about which everything is flipped the *symmetry axis* of the figure.

This gives a precise mathematical characterization of bilateral symmetry: a shape is bilaterally symmetric if there exists some reflection that leaves it *invariant* – that is, unchanged in appearance. Admittedly this formal description of symmetry doesn't capture its aesthetic aspects very well – a horribly ugly shape can be bilaterally symmetric in the mathematical sense; imperfections that destroy mathematical symmetry may add aesthetic value – but we never intended to reduce art to mathematics. Every shape, incidentally, has a 'trivial' symmetry – leave it alone. Even though it's trivial, it's important, so it gets a special name: the *identity*. Bilaterally symmetric shapes have two symmetries: the identity, and a flip.

The Greedy Starfish

There are other types of symmetric shape, for example a five-limbed starfish. Again, we think of an ideal starfish, all of whose limbs are identical. What are the corresponding transformations? The most obvious is the one that makes all the limbs 'click one space on': a *rotation* through an angle of 72° (one-fifth of a full turn) (Figure 2.2a). If you leave a starfish lying on a table and while your back is turned somebody rotates it through 72°, you won't know anything has happened. But if instead they rotate it through 45°, say, then you'll spot a change in its orientation. There are precisely five different angles through which an ideal starfish can be rotated without the change being detectable: 0°, 72°, 144°, 216°, and 288° – the integer multiples of one-fifth of a turn. Informally, we say that the starfish has *five-fold rotational symmetry*. Rotations in the plane don't have an axis, but they do have a special point, the centre of rotation, which doesn't move at all. As its name suggests, in this case it's the point right in the middle of the starfish.

These rotations are the most noticeable symmetries of a starfish, but it actually has more, because a starfish in a mirror still looks like a starfish. Each limb is bilaterally symmetric, and the limb's symmetry axis passes through the centre of rotational symmetry, so, like us, the starfish has bilateral symmetry. Indeed the greedy beast has five different bilateral symmetries, because it has five arms and each has its own symmetry axis. There are five distinct reflections that leave a starfish invariant, and their axes are inclined at angles of 72° to each other (Figure 2.2b). In total, a starfish has exactly ten symmetries. (For similar reasons, the milk-drop splash discussed in chapter 1 has 48 symmetries: 24 rotations, and 24 reflections.)

This insight, that objects possess not just symmetry, but symme*tries*, transformations that leave them invariant, means that not only are mathematicians in possession of a qualitative description of symmetry; it also has a quantitative aspect. We can *prove* that the symmetry of a starfish is *different* from that of a human being; for starfish have ten symmetries and we have only two. Well, yes, of course – that's hardly an impressive achievement, on its own. But the ability to tackle symmetry systematically, to compare the symmetries

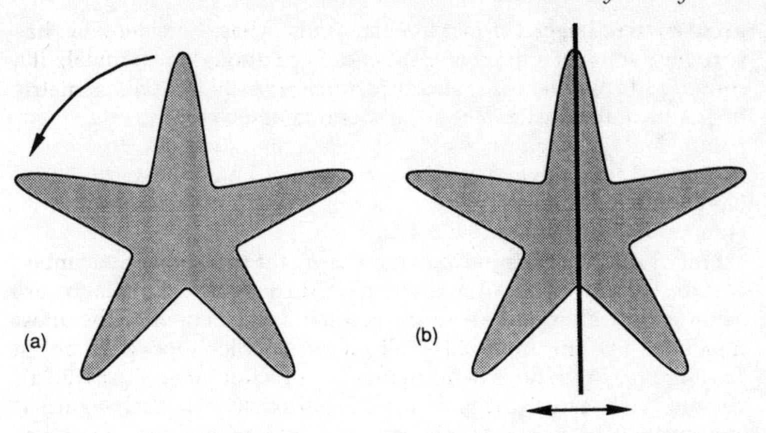

Figure 2.2 Symmetries of a perfect starfish. (a) Rotation. (b) Reflection

of different objects – that's very powerful indeed. For example, we shall shortly see how a young Frenchman proved that equations of the fifth degree can't be solved by a formula – purely because they have the wrong kind of symmetry.

What is a Transformation?

In order to move ahead, it may at this point be best to take a step backwards, and begin by explaining what a transformation is. In fact, we'll take a further step backwards and explain what *explanation* is, as far as this book is concerned. We need to agree on some ground rules. Otherwise, while we're busy trying to describe, in necessarily vague and flowery language, what the idea behind some mathematical concept is, you'll be sitting there imagining that mathematics is vague and flowery, and wondering what happened to its legendary precision. The main ground rule is that, precisely because we *don't* want to burden you with technical details, we will have to discuss concepts by example and analogy, rather than by giving the actual mathematical definitions that you'd find in the textbooks.

For instance, there's a rigorous and precise mathematical definition of transformation, based upon set theory. If you want the flavour of it, it starts something like this: 'A function is a subset of the cartesian product of two sets with the following three properties ...' See, it's so technical it doesn't even use the word 'transformation'! We want to avoid getting bogged down in that sort of mess, because you need several years of training before you can appreciate how wonderfully simple and enlightening it all is. (Really, it is. Ask any mathematician. In mathematics, as in Zen, the attainment of enlightenment is a painful and lengthy process.) If the full technicalities of mathematical concepts were easily grasped, there'd have been no need for Weyl to have written his masterpiece, or for us to have embarked upon the volume that you hold in your hand.

End of diatribe. Assuming your agreement to these ground rules, we can tell you what a transformation is. Basically, it's a recipe for moving things around. To each possible thing, the transformation associates a second thing, its *image*. For example, supose the things are numbers, and the transformation is 'associate to each number its negative'. Then the image of a number, say 7, is its negative, −7. Nothing to it! Notice that we said that a transformation is a recipe: it is a *process* that determines the image, and it is not the image itself. The process here can be expressed algebraically as the rule 'multiply by −1', and that's what determines the transformation; but −7 is just a number. You 'know' a transformation when you can work out the image of *any* initial thing.

For mental vividness, it's convenient to think of a symmetry transformation as a motion: pick the object up, move it about, and put it back down 'on top of itself' – into the same space that it started from. It's customary to think of numbers as being strung out along a line, with the negative ones to the left and the positive ones to the right. The algebraic description of the above transformation is 'multiply by −1'. That swaps positive and negative numbers, so it interchanges the left and right halves of the number line: geometrically speaking, it's a reflection. (You may prefer to think of it as a rotation: just pivot the whole line through 180° about its centre. We'll return to this point later.) This duality between algebra and geometry was discovered by René Descartes: every geometric object has an algebraic description, every algebraic formula determines a geometric object. Humans tend to use the algebraic version for calculation, and the geometric one for imagination.

There's a further important thing to understand about this kind of imagery: the only information that matters is the correspondence between initial points and their images under the 'motion'. Where they go in between may be helpful in visualizing this correspondence, but it's not part of the meaning of 'transformation'. You can keep the starfish on the table while you gently rotate it, or you can pick it up, toss it up and down in your hands, and finally replace it in a rotated position. The starfish might well prefer the first sequence of events, but the mathematics is indifferent: the sole important feature is that when the starfish gets back into position each point of it has rotated through an angle of 72°.

Rigid Motions

The main kinds of transformation that we are thinking about are *rigid motions* of space. Here the 'things' are the points of space – usually ordinary three-dimensional space, but the dimension might be any number. 'Rigid' means that points end up the same distance apart as they were to begin with; that is, the images of two points are the same distance apart as the points themselves. The word 'motion', given what we've just said, is just a convenient term to describe a process, related to everyday experience, that produces the correct end result.

Let's begin with a one-dimensional space – a line. What are the possible rigid motions? Basically, we have to pick the line up, without bending or stretching or squashing it, and lay it back down upon itself. How can we do that? Find a ruler, or a pencil, which are good enough approximations to a line provided you imagine them to be extended indefinitely, and experiment. You could, for instance, slide the line along, either to left or right. If *one* point moves, say, five units to the left, then rigidity implies that *every* point must move five units to the left. So the recipe, for any point, is 'move five units to the left'. Such a transformation is called a *translation* (Figure 2.3a).

What else could you do? You could *flip the line over*, interchanging left and right, and then replace it. That's best thought of as a reflection (Figure 2.3 b). You may have read Edwin A. Abbott's mathematical fantasy *Flatland*, which is mostly about polygonal creatures that inhabit a plane. However, in one chapter the protagonist, A. Square by name, takes a trip to Lineland, a one-dimensional world inhabited by pointike beasts. If you were a Pointy Thing living in Lineland, and you looked at yourself in a point mirror, then the part of the line that lay behind you would appear to be behind your image in the mirror – that is, flipped to the opposite direction. (Actually, your image would block your view of it, unless you were transparent, but that's being pedantic.)

It's not hard to see that these are the *only* possible rigid motions of a line; translate through some distance, or reflect in some particular point. Consider the possibilities. If you place the line back where it was without flipping it end for end, then all you can do is slide it (a translation); if you flip it end for end then there is some 'pivot' point

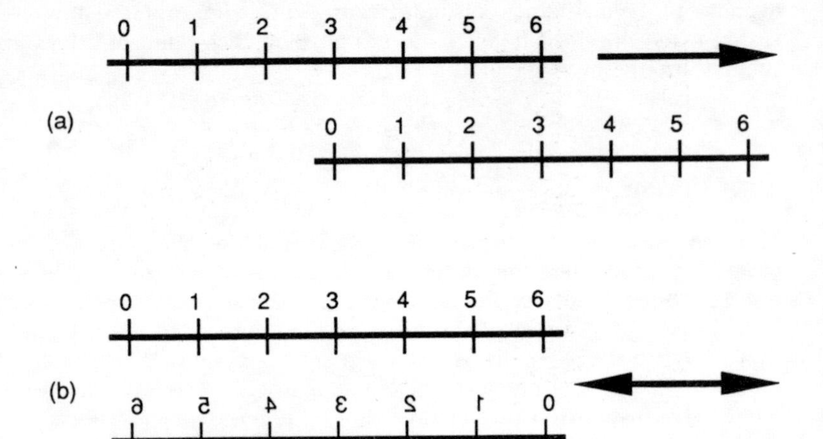

Figure 2.3 Symmetries of a line. (a) Translation. (b) Reflection

that doesn't move at all, and that's where Pointy Thing puts her mirror to get a reflection.

What about two dimensions? Now you have to pick up a whole plane and put it back again, and you should experiment with a sheet of paper. There are motions analogous to translations: slide the entire plane some chosen distance in some chosen direction (Figure 2.4a). 'Two units North and six West', for example. Notice that when using this kind of description we have to specify two distances, one in each of two mutually perpendicular directions: that's because the plane is a two-dimensional space. Indeed it's pretty much what mathematicians *mean* when they say 'two-dimensional space': it has a coordinate system in which two numbers specify each point.

As well as sliding the plane, you can also turn it a bit, through an angle (Figure 2.4b). The simplest way to do this is to choose some particular point, and insist that it remain fixed. Take a piece of paper, put it on a board, and stick a push-pin into it somewhere. Now turn it round, keeping it flat. That's a *rotation*, and the pin is its *centre*. It may seem that you get something new if you slide the plane as well as turning it, but in fact any motion of that kind also has a fixed point somewhere; it's just a rotation about a different point.

Translations and rotations keep the plane the same side up. But we can also flip it over, which again yields reflections (Figure 2.4c). This time, they are reflections in some *line*. Draw a horizontal line across

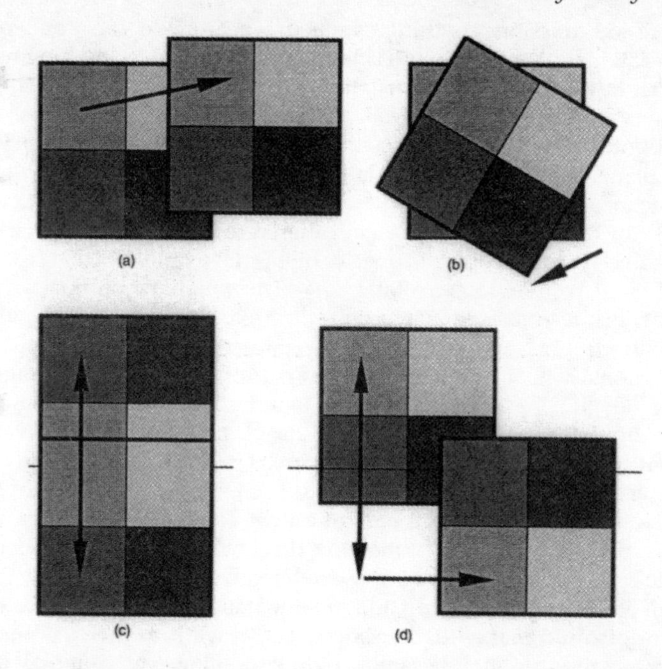

Figure 2.4 Symmetries of the plane. (a) Translation. (b) Rotation. (c) Reflection. (d) Glide reflection

the exact middle of your sheet of paper. Repeat the line on the back, in the same place, so that you can see it from either side. Flip the paper over, top to bottom, so that it ends up in the same position as before. Observe that the line stays where it is; indeed *every point on the line* stays where it is, and *every other point moves*. Points originally above the line end below; points below end up above. The line, and only the line, stays fixed.

To see that this really *is* a reflection (despite what you did to the paper, which was a rotation in three-dimensional space) imagine yourself as A. Square, living in Flatland, and looking at himself in a mirror that runs along the line you drew. A point three units below the mirror, for example, will have an image that is three units above it – which is exactly where it goes when you flip the paper over. Find a pocket mirror, set it up along the line, and check.

There are lots of reflections in the plane, because you can use any line whatsoever as a mirror. They don't have to run vertically or horizontally: they might be at a slope.

Are we missing anything? Indeed we are: we can combine a translation with a reflection. Usually, this doesn't give anything really novel; but it does if the translation is parallel to the mirror. Then it yields a motion called a *glide reflection* (Figure 2.4d), which we encountered earlier in connection with the flow of air past a truck. In the plane, that's the lot: translation, rotations, reflections, and glide reflections.

On to three dimensions. Again there are translations, which this time need three numbers to specify them ('Two units North, six West, and five Up.') There are rotations, which now have not just a fixed point, but also a fixed *axis*: a line through that point, acting like a spindle around which the rest of space revolves. And there are reflections, which now happen in mirror planes, familiar enough in everyday life. Yes, and there are glide reflections too. Anything else? Of course, there's *always* something else ... If you combine a rotation about some axis with a translation parallel to that axis, then you get a helical motion known as a *screw* (Figure 2.5), which we've already met when discussing the spiral bark of a tree. That's the lot.

Reflections, and transformations that include reflections such as glide reflections, have a slightly paradoxical air. You can't apply them to an object by picking it up and moving it around. If you could, then factories could make just left boots, and leave it to the customer to turn one round until it became a right boot. However, you may have noticed that when discussing *reflection* in the plane, we suggested you

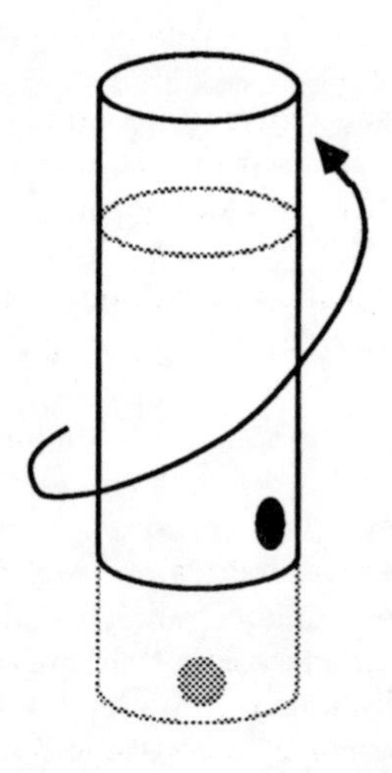

Figure 2.5 A screw symmetry

rotate a piece of paper. Did we make a mistake? We admit to being capable of it; but the answer here is 'no'. The reason is subtle and intriguing. The rotation that we used to turn the paper over *lifts it out of its original plane*. It's a rotation in three-dimensional space. When we put the paper back again, the *end result* is exactly that of a reflection in two-dimensional space. (You must imagine mathematical paper of zero thickness: points are *in* it, not on either side of it.) We've already insisted that a transformation be judged by its end effect, not by what happens in between.

Reflections in three-dimensional space at first seem different: this time you can't find a rotation that produces the same result. Immanuel Kant considered this to be of deep philosophical significance. Michael Flanders and Donald Swann wrote a less deep song, *Misalliance*, based upon an impossible affair of the heart between two different species of creeper:

> To the Honeysuckle's parents it came as a shock.
> 'The bindweeds', they cried, 'are inferior stock.
> They're uncultivated, of breeding bereft.
> We twine to the right, and they twine to the left!'

But appearances are deceptive. A Flatlander can't rotate a plane to obtain the same effect as a reflection; but a Spacelander can. Similarly a Spacelander can't turn a solid into its mirror-image by a rotation – but it can be done by a creature with a four-dimensional space at its disposal. (Such creatures do exist. What are they? See below.) In four-dimensional space it's possible to take a three-dimensional object, and – like the sheet of paper – rotate it through the fourth dimension to end up with its reflection. Indeed, reflections in any dimension of space can be considered as rotations in a space one dimension higher.

Oh, yes – the creature with a four-dimensional space at its disposal is a mathematician, of course.

The Group Concept

The understanding that symmetries are best viewed as transformations arose when mathematicians realized that the set of symmetries of an object is not just an arbitrary collection of transformations, but has a beautiful internal structure.

Let's return to the starfish as an example. We now know that there are ten distinct transformations that leave a starfish invariant: five

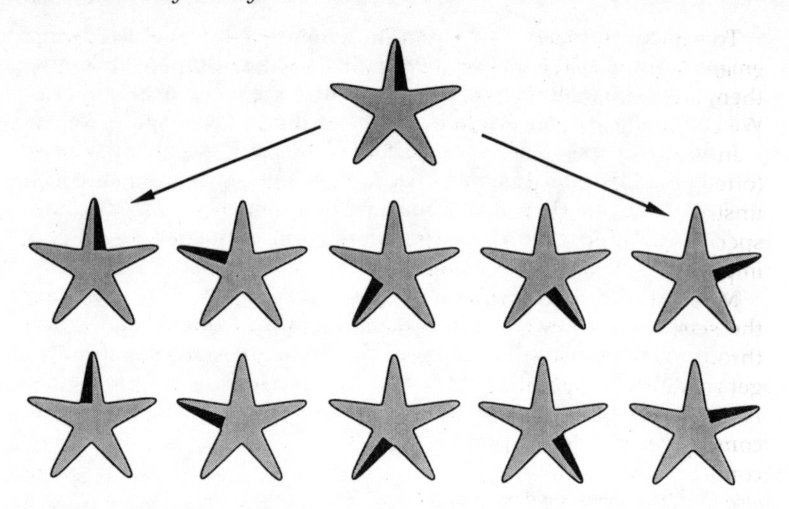

Figure 2.6 The ten symmetries of a starfish. The shaded region is marked to represent the motion. The starfish at the top is for reference: the others show its images under the ten symmetry transformations. The first transformation is the identity: leave it alone

rotations (one trivial) and five reflections (Figure 2.6). Suppose that two of these transformations are applied, one after the other. Each leaves the starfish apparently unchanged, so the final result also leaves it apparently unchanged. That is, the result of performing two successive symmetry transformations must be another symmetry transformation. For example, let's rotate the starfish by one-fifth of a turn, and then by a further two-fifths of a turn. What it the result? Clearly the combination of the two rotations has exactly the same effect as rotating it three-fifths of a turn. But that's just another symmetry of the starfish! With a little thought we can see that this property holds in considerable generality: for any shape whatsover, any two of its symmetries, when combined together, must yield another symmetry. The reasoning is straightforward: a symmetry is a transformation that leaves the object invariant – that is, it looks the same after applying the symmetry as it did to begin with, it is replaced into the same space that it started from. Take the object, and apply the first symmetry: it looks unchanged. Now apply the second symmetry: it *still* looks unchanged. So after applying them *both*, it looks unchanged. But that means that the combination of the two symmetries is itself a symmetry. If you leave something unchanged *twice*, you leave it unchanged.

To express this fact, we say that the symmetries of an object form a *group*. A group is a *closed* system of transformations: whenever two of them are combined, the result is another member of the same group. We call this particular group the *symmetry group* of the object.

In ordinary language, a group is just a bunch, a number of things (often people) lumped together. In mathematics the word for such an unstructured collection is 'set'; and the word 'group' has a more specific meaning – a set of transformation *plus* the structure that is imposed upon it by the possibility of combining its members in pairs.

Most sets of transformations don't form a group. For example, take the set consisting of just two transformations of a starfish, rotation through one- or two-fifths of a turn *only*. When we combine these, to get rotation through three-fifths of a turn, then we go outside the set. A group, however, must be 'closed': you *can't* go outside it by combining transformations that are already in it. So, if a *group* contains rotation through one- and two-fifths of a turn, then it must *also* contain rotation through three-fifths. Moreover, by combining a one-fifth turn with the 'new' three-fifths turn we find that it must also contain rotation through four-fifths of a turn. Finished? Not yet: if we combine rotation through four-fifths of a turn with rotation through one-fifth of a turn we see that it must also contain rotation through five-fifths of a turn – which has the same effect as a rotation through zero, the trivial transformation that leaves every point where it started.

Indeed, if all we know to begin with is that some group contains just *one* transformation, rotation through one-fifth of a turn, then we can combine that with *itself* to deduce that the group also contains rotation through two-fifths of a turn, and then as before through three-, four-, and five-fifths (the same as zero in Starfishland). 'Give them an inch and they'll take an ell', says the proverb. Groups are just like that – if you give them an inch then they'll combine it with itself 44 times in succession to get 45 inches, which is how long an ell is.

Algebra in Starfishland

Groups are thus very special sets of transformations, so the fact that the symmetries of an object form a group is a significant one. However, it's such a simple and 'obvious' fact that for ages nobody even noticed it; and even when they did, it took mathematicians a while to appreciate just how significant this simple observation really is. It leads to a natural and elegant 'algebra' of symmetry, known as

Group Theory. Here's a tiny hint of what's involved, enough for you to see that 'algebra' is a sensible word to use. We've seen that rotation through one-fifth of a turn, combined with rotation through two-fifths of a turn, yields rotation through three-fifths of a turn. Symbolically, we can write this (in units of 'one turn') as $\frac{1}{5} + \frac{2}{5} = \frac{3}{5}$, a natural enough equation. The mathematics of symmetry groups is not *quite* as simple as this, however. Imagine rotating three-fifths of a turn and then two-fifths. The result is a full turn; but that really does leave every point of the starfish in exactly the same place that it started. If we concentrate only on where points end up, and not on how they got there, this is the same as 'no rotation'. In other words, $\frac{3}{5} + \frac{2}{5} = 0$ in the world of starfish symmetries!

For fun, and to make sure you've grasped the point, cut out a paper starfish, and see what happens when you combine two of its reflectional symmetries. Do you get another reflection? If not, what do you get? Is there any easy way to see what the answer to the first question must be?

The mathematics of symmetry groups becomes considerably more complicated when the transformations take place in three-dimensional space. For example, we'll mention in chapter 4 the discovery that there are exactly 230 distinct types of symmetry groups of crystals. Ideally, crystallographers should be intimately acquainted with the individual mathematical characteristics of each one!

The Gambling Scholar

Insights of this simplicity and depth don't arise because people sit down saying to themselves 'what is symmetry?' They happen because hints of the general idea emerge from important problems. One of the first mathematical theorems about symmetry was discovered by the ancient Greeks – it's the culmination of Euclid's multi-volume textbook of geometry, the *Elements*. Euclid proved that there exist precisely five regular solids (Figure 2.7). A solid is regular if each face is an indentical regular polygon and if the arrangement of polygons at each vertex is the same. A regular polygon, in turn, is a figure composed of straight lines of equal length, joined at equal angles. The five regular solids are the:

- tetrahedron, with four triangular faces;
- cube, with six square faces;
- octahedron, with eight triangular faces;

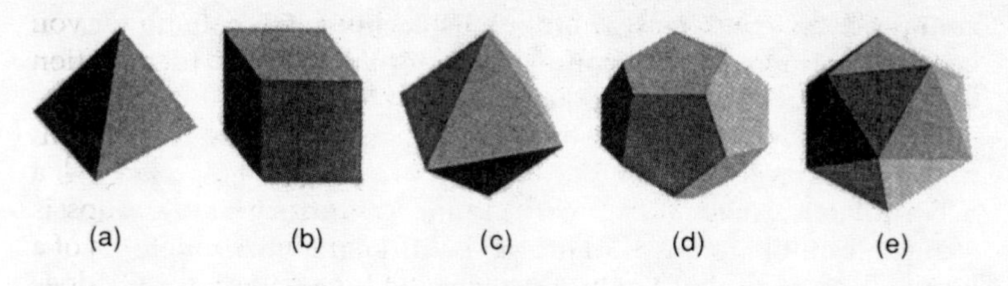

Figure 2.7 The five regular solids. (a) Tetrahedron. (b) Cube. (c) Octahedron. (d) Dodecahedron. (e) Icosahedron

- dodecahedron, with twelve pentagonal faces;
- icosahedron, with twenty triangular faces.

The regular solids were known to Plato, and indeed they're often called the Platonic solids. Their high degree of symmetry appealed to Plato's philosophical tendencies: a Geometer God couldn't fail to make fundamental use of such elegant forms. But there's no evidence that either Plato or Euclid possessed a mathematical formalization of the concept of symmetry: to them the appeal of the regular solids seems to have been primarily aesthetic.

Ironically, the source of the group concept wasn't geometry, but algebra: the solution of equations. Most of us were taught about quadratic equations at school – they involve not just an unknown quantity, but also its square – and those who were may still recall that there's a formula for the answer. The process that provides that provides that formula, though not the formula itself, was known to the ancient Babylonians in 1600 BC, an eloquent testimony to the speed with which mathematical concepts permeate human culture. Beyond quadratic equations come cubics, involving the cube of the unknown in addition to its square and first power. A similar but more complicated formula to solve those was discovered by two Renaissance mathematicians, Scipio del Ferro and Niccolo Fontana (nicknamed Tartaglia, 'The Stammerer'). At that time mathematics was a competitive event, the pecking-order and contingent patronage by the nobility being established by public displays of computational prowess – such as solving equations set by one's opponent – and at first they kept their methods secret. However, Fontana revealed them to Girolamo Cardano, the 'gambling scholar' and one of the most colourful figures in the history of science, swearing him to secrecy. Cardano promptly published them (albeit giving due credit to the discoverer) in his algebra text *Ars Magna* of 1545. The book also contained an even more horrendous formula for solving quartic equations – involving the fourth power of the unknown – found by Cardano's student Ludovico Ferrari.

On to the quintic – fifth powers – and an even more horrendous formula?

Fortunately not. For nearly three centuries mathematicians struggled, until around 1770 Joseph-Louis Lagrange began to see the light. He showed that the formulas for solving quadratic, cubic, and quartic equations all reduce to a single general trick, and that that trick *fails* on the quintic. Of course, there might be *another* trick that works; but nobody could find one, and Lagrange's idea pinpointed the nature of the problem. The trick, said Lagrange, is to study rearrangements, or *permutations*, of the putative solutions of the equation. For example, the six solutions *a*, *b*, *c*, of a cubic can in general be permuted in six ways: *abc*, *acb*, *bac*, *bca*, *cab*, *cba*. In 1824, by pursuing this insight, Niels Hendrik Abel was able to prove conclusively that the long-sought formula *did not exist*. The formula for the quintic was a mathematical mare's nest.

Femme Fatale

Abel's proof was complicated, and didn't fully penetrate to the heart of the problem. That honour fell to a young Frenchman, Évariste Galois, who was born in 1811, died in 1832, and spent much of the time between getting into trouble. Galois developed a general theory of permutations of the solutions to equations, concentrating on those permutations that preserve all algebraic relations between the solutions. He noted a remarkable property of these systems of permutations: if two of them are performed in turn, then the result is always another permutation in the same system. He called such a system a *group* of permutations. As the culmination of his work he showed that equations that can be solved by a formula must have groups of a particular type, and that the quintic equation *has the wrong sort of group*. It's as simple as that.

Galois was a colourful, ultimately tragic, figure. His work wasn't appreciated during his lifetime, and the mathematical establishment ignored him. He died in a futile duel – pistols at 25 paces – over a woman, which may have been a set-up by his political enemies. Galois wrote that the young lady, until recently known only as the mysterious Stéphanie D., was an 'infamous coquette'. However, some detective work on Galois' manuscripts by the historian Carlos Infantozzi has unveiled the *femme fatale* as Stéphanie-Felicie Poterin du Motel, the entirely respectable daugher of a doctor who was one of Galois' neighbours. It seems likely that Galois' attentions were not welcome, and that (typically) he was carried away by his emotions and reacted badly.

Be that as it may, Galois had extracted the key idea of a group from the algebraic intricacies of the quintic equation; and he knew full well how important it was. It took the rest of mathematics a couple of decades to come to the same conclusion, and it might well have taken longer were it not for the efforts of Joseph Liouville, the first to realize what gems lay buried in Galois' writings.

Permutations of solutions of equations may seem a far cry from transformations of geometrical shapes. However, there's a very real sense in which Galois' permutation groups may be thought of as the symmetry groups of the equations. First, a permutation is itself a transformation: not of space, but of the collection of solutions. It is a rearrangement, a way of 'picking them up and replacing them'. Now, in order to be a symmetry of a geometrical object, a transformation must leave its shape unchanged. So what, in the theory of equations, is the analogue of shape? Well, the shape of an object is determined by the distances between its component points, and a symmetry is a rigid motion, preserving those distances. Moreover, distance is the basic geometric relation between points. Analogously, symmetries of an equation ought to preserve the basic *algebraic* relations between its solutions, and that's exactly how Galois defined his groups. It's unlikely that the analogy was apparent to Galois, however, because groups of symmetries in geometry were still some way off. It was certainly well established by the time of Hermann Weyl, who – using a piece of jargon which we've not mentioned but which he immediately explains – wrote:

> *Whenever you have to do with a structure endowed entity ... try to determine its group of automorphisms*, the group of those ... transformations which leave all structural relations undisturbed.

The italics are Weyl's: he clearly thought the advice was important.

Geometry Is Symmetry

Until now, we've been viewing symmetry as a kind of accident of geometry. In typical fashion, mathematicians turned everything upside down, and began to view geometry, or more accurately, geometr*ies*, as a consequence of symmetry. Geometries? Isn't one enough? Certainly not! In Euclid's day there *was* only one geometry – Euclid's, naturally – but by the 1870s geometry had proliferated into an unruly mob: Euclidean and non-Euclidean geometry, projective geometry, affine geometry, conformal geometry, inversive geometry, differential geometry, and the first flushes of topology. A believer in a Geometer God would have had to be a pantheist.

It was Felix Klein, a German mathematician at the University of Göttingen, who made symmetries fundamental and geometries subsidiary. In the encyclopaedic Germanic tradition, he wanted to impose order on the chaos of geometry. But instead of just cataloguing all the possibilities, he introduced an important new element of mathematical structure. In 1872, at the University of Erlangen, he gave a lecture now known as the Erlangen Programme. The essence of Klein's programme is that geometry is group theory. The groups are formed from the transformations that leave the basic geometric notions invariant – but this relationship can be inverted, so that, in Klein's words, 'geometrical properties are characterized by their invariance under a group of transformations'. Each type of geometry has its own group; but within the framework of that group, each geometry proceeds along analogous lines. Group theory provides the common ground that links different geometries. Let's look at four examples.

In Euclidean geometry, for example, the basic notions are distances and angles. The transformations that preserve distances and angles are precisely the rigid motions. Effectively, Klein's idea is to reverse this argument, take the group of rigid motions as the basic object, and deduce the geometry. So a legitimate geometric concept, in Euclidean geometry, is anything that remains unchanged after a rigid motion. 'Right-angled triangle', for example, is such a concept; but 'horizontal' is not, because lines can be tilted by rigid motions. Euclid's obsession with congruent triangles as a method of proof – whose opacity has baffled and infuriated generations of schoolchildren – now becomes transparent, for triangles are congruent precisely when one can be placed on top of the other by a rigid motion. Euclid used them to play the same role as the transformations favoured by Klein.

Non-Euclidean geometry is also based on distances and angles, but they don't behave exactly like their Euclidean counterparts. Non-Euclidean geometries are designed to violate Euclid's axiom of parallels: through any given point there exists a unique line parallel to any given line. In one variety, elliptic geometry, parallels don't exist at all; in the other, hyperbolic geometry, parallels come in infinite bunches. Each type of non-Euclidean geometry has its own group of rigid motions – motions that preserve its own peculiar notion of distance. To give the flavour of these strange (but highly important) geometries, Figure 2.8 shows an Escher lithograph based upon hyperbolic geometry. Although the angels and devils appear to shrink as they move towards the edge of the circle, that's true only for the usual Euclidean notion of distance. In the appropriate distance for hyperbolic geometry, all the angels and all the devils are identical: they form a tiling of the hyperbolic plane. It's clear that this one has a lot of symmetry.

In projective geometry, the permitted transformations are projections – much as a film is projected on to a screen. Projections don't preserve distances (Harrison Ford on screen is much larger than his image on the strip of film in the projector, and that in turn is smaller than his actual size when the film was shot), so distances are not a valid concept in projective geometry. 'Elliptical' is, however, because any projection of an ellipse is another ellipse.

Figure 2.8 Escher's Circle Limit IV *of 1960 represents the geometry of the hyperbolic plane, from which point of view all the devils have the same shape and size, as do all their complementary angels*

Topology allows a much wider range of transformations: continuous ones, which can bend, stretch, or squash space, but not tear or cut it. 'Elliptical' is not a topological concept, because an ellipse can be distorted continuously to yield a square, or a triangle. 'Knotted' is a typical topological concept: you can't undo a knot in a closed loop by stretching or bending it.

See? Different groups place different emphases, but each leads to a self-consistent style of geometrical analysis.

As well as bringing order to chaos, Klein's unification revealed unexpected connections between different geometries. Some geometries turn out to have groups that are the same as those of other geometries, perhaps thinly disguised. That means that the two geometries are really equivalent to each other: a mechanical procedure can convert the true theorems of one geometry into true theorems of the other. In a similar way, geometries based upon larger groups are more general than those based upon smaller ones; and any valid theorem in the geometry with the big group is automatically valid in that with the smaller one. For example, theorems of projective geometry are automatically valid theorems of Euclidean geometry. So there's a kind of hierarchy of geometries, in place of a disordered muddle.

Klein's view was very influential, not just because it unified the vast range of geometries, but because mathematicians of his period were finding that more and more of their problems hinged around transformations and groups. Henri Poincaré said that 'The theory of groups is, as it were, the whole of mathematics stripped of its matter and reduced to "pure form." ' It was a bit of an exaggeration, but it contained more than a grain of truth: groups certainly underpinned most of the mathematics *that interested Poincaré*, and he was one of the broadest-minded of them all.

After Klein used it to bring the unruly mob of geometries under control, the subject of group theory really took off. Arthur Cayley introduced a more abstract setting for group theory, in which the role of the transformations was de-emphasized, and the manner in which they combined became paramount. A much more general notion of 'group', a system not necessarily composed of transformations at all, emerged. We won't go into that, because the groups that interest us in the study of symmetry-breaking always arise through transformations. Nor shall we pursue the history of group theory any further, fascinating though it is, because that would divert us from our main objectives. Instead, we'll introduce the groups that are most important for the rest of this book: groups of rigid motions in two- and three-dimensional space.

A Gaggle of Groups

Let's take a look at some of the basic symmetry groups, to get some kind of feel for what goes on. We're not after an exhaustive catalogue – there being, for instance, 230 different crystal symmetry groups in three dimensions, that seems a wise decision.

The simplest groups of all are called *cyclic* groups. A good example of a shape with a cyclic symmetry group is the Isle of Man's 'running legs' symbol (Figure 2.9). It consists of three legs, chasing each other round and round like spokes on a wheel. The symmetries of this shape are rotations, and there are three of them: rotation by one third of a turn, by two thirds of a turn, and by zero (the trivial symmetry). The 'algebra' of these rotations is straightforward: if you combine two rotations you add them, but then you must remove any complete turns that may have accumulated. For example, in units of one turn, $\frac{2}{3} + \frac{2}{3} = \frac{4}{3}$, but since this is greater than a full turn we deduct one turn, $\frac{3}{3}$, to leave $\frac{1}{3}$.

A similar shape, but with n equally spaced legs, has a symmetry group consisting of rotations through various multiples of $\frac{1}{n}$ of a turn.

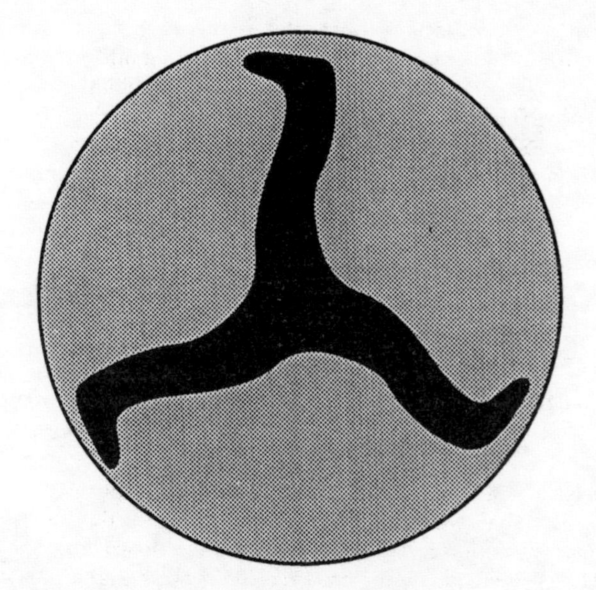

Figure 2.9 The Isle of Man's running legs are symmetric under rotations through multiples of 120°

There are precisely n such rotations, and they form the *cyclic group of order n*, for which mathematicians use the symbol Z_n. Unlike the symmetry group of the starfish, cyclic groups don't contain any reflections. Think of a starfish with feet on the ends of its arms, like the running legs symbol has: its symmetry group is then cyclic of order 5, because the asymmetric feet remove the possibility of reflections.

The symmetry group of the starfish is an example of a *dihedral group*. The simplest way to think of the dihedral group is as the symmetry group of a regular n-sided polygon. As well as the n rotations of the cyclic group, there are a further n reflections in the polygon's symmetry axes (Figure 2.10). So the group contains $2n$ symmetry transformations altogether: half rotations, half reflections. The notation for it is D_n.

The rotational symmetries leave the polygon facing the same way up; the reflections effectively flip it over. So what happens if you combine two reflections? It flips over twice, so it ends the same way up. But symmetries that leave it the same way up must be rotations. So, *without* making any detailed calculations, we see that in D_n the combination of two reflections is always a rotation. Group theory

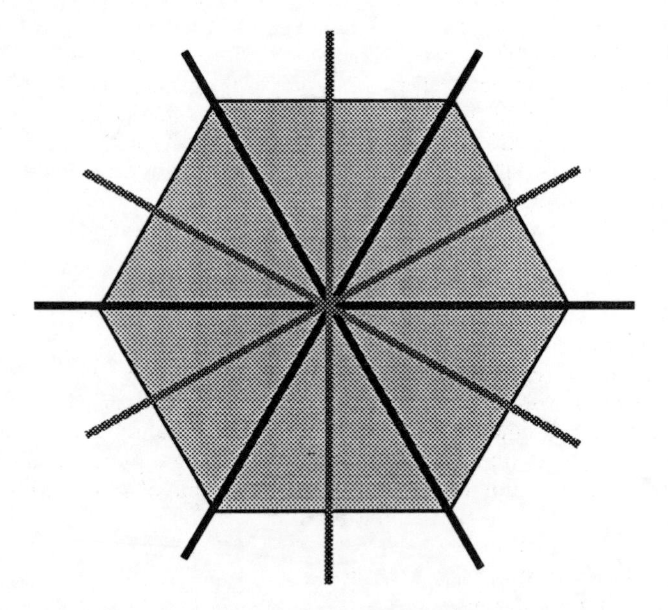

Figure 2.10 Symmetry axes of a regular n*-sided polygon with $n = 6$*

derives a lot of its power from this kind of simple argument. Precisely which rotation depends upon the angle between the two mirror axes, and the order in which the reflections are performed. You have to get your hands dirty eventually!

The groups Z_n and D_n are said to be *finite*, because they contain a finite number of transformations. Buildings with polygonal symmetry were common during the Italian renaissance, and Leonardo da Vinci investigated all the possible symmetries of a main building with adjacent chapels – effectively proving a theorem in group theory: the *only* finite groups of transformations of the plane are Z_n and D_n. This is one of the earliest instances of a theme that will recur throughout this book: the ability of mathematical arguments to determine all possible types of symmetry group of some particular kind. A still earlier example is the classification of the five regular polyhedra, known to Plato, to which we shall return in chapter 4.

A circle can be thought of as a regular polygon with infinitely many sides. Its symmetry group consists of *all* rotations about a chosen centre, together with all reflections in axes that pass through that centre. Again, combining any two reflections yields a rotation. The symbol for the group is O(2), the *orthogonal group in two dimensions*. The rotations alone form a smaller group SO(2), the *special orthogonal group in two dimensions*.

Let's move on to three dimensions. Here we find several new groups. The symmetry group of a sphere, for example, consists of all rotations about its centre, and all reflections in planes that pass through that centre. Together these form O(3), the *orthogonal group in three dimensions*. The rotations alone form a smaller group SO(3), the *special orthogonal group in three dimensions*.

The five regular solids have especially interesting symmetry groups. Let's start with a cube, because that's the most familiar shape. A cube has eight vertices, and three edges meet at each vertex. Now, we can rotate any given vertex to any position we wish; and having done so, we can rotate the cube, leaving that vertex fixed, to cycle the three edges that enter that vertex. So the cube has $8 \times 3 = 24$ rotational symmetries. Moreover, it has at least one reflectional symmetry, and combining that with the 24 rotations we get a further 24 reflectional symmetries, making a grand total of 48.

By similar arguments, we can build up a table of symmetries, like this:

- tetrahedron, 24 symmetries;
- cube, 48 symmetries;
- octahedron, 48 symmetries;

- dodecahedron, 120 symmetries;
- icosahedron, 120 symmetries.

The cube and octahedron both have the same number of symmetries (48), and so do the dodecahedron and the icosahedron (120). The reason for this is a phenomenon known as duality. If you place a dot at the centre of each face of a cube, you get six dots, forming the vertices of a regular octahedron, its *dual*. Every symmetry of the cube is also a symmetry of the dual octahedron; every symmetry of the dual octahedron is also a symmetry of the cube. So the two solids have the same group of symmetries. Similarly, if you place dots at the centre of the faces of a dodecahedron, you obtain a dual isocahedron, and again the two solids have the same group (Figure 2.11).

What about the tetrahedron? Dots at the centres of its faces yield another tetrahedron. It's self-dual, and we get nothing new. So instead of five different groups, there are just three, usually called the *tetrahedral*, *octahedral*, and *icosahedral* groups. Felix Klein made a systematic study of their applications to functions of a complex variable, and discovered connections between the icosahedral group and Galois' theory of the quintic equation. We are paddling at the edge of deep waters.

These are some of the main symmetry groups of three-dimensional shapes. There are others, which are a little more subtle to describe, but we're not after a complete list here. Instead, let's take a brief look at the two-dimensional analogue of crystal structure. You'll see one such pattern on most bathroom walls: square tiles. An infinite plane covered with square tiles has two quite distinct types of symmetry. First, you can select a particular tile, and apply symmetries that leave that tile in the same place. You get eight symmetries for each tile, forming a dihedral group D_4, just like the symmetries of a single square. But there are also symmetries that move a given tile to some

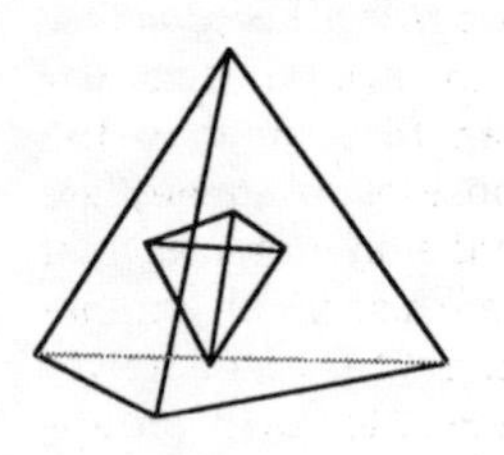
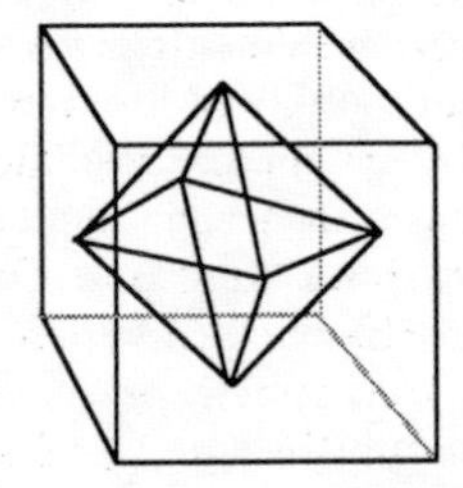
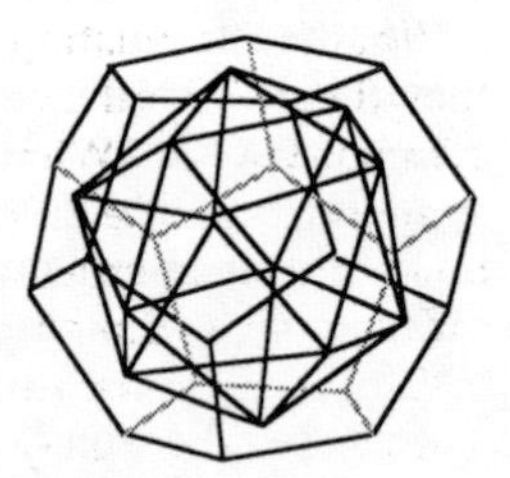

Figure 2.11 Duality between pairs of regular solids

other tile. If the orientation doesn't change, these are translations, through a whole number of tiles horizontally and a whole number of tiles vertically. To change orientation as well, you can first apply a translation to switch tiles, and then rotate the final tile. Therefore every symmetry of the entire pattern of tiles is a combination of these two types. The translations are called *lattice symmetries* and the individual D_4s are *point groups*. We'll have more to say about this kind of group when we talk about crystals in chapter 4, and about symmetric chaos in chapter 9.

Symmetry-breaking Through the Eyes of a Group-theorist

When Weyl first introduces groups in his book, he does so from an abstract point of view, and then asks: 'What has all this to do with symmetry?' His answer is that group theory 'provides the adequate mathematical language to define it'. If so, then group theory should also provide the adequate mathematical language to define symmetry-*breaking*. Let's see how.

Think about the buckling ping-pong ball. What happens to the symmetry group of the ball as it buckles? Before buckling, the ball is a perfect sphere. Its symmetry group consists of all rigid motions that fix its centre. These are rotations, through any angle, about any axis through the centre, or reflections in any plane through the centre. We've just seen that the totality of such transformations forms the group known as O(3), the orthogonal group in three dimensions.

After buckling, there's a preferred axis, and the buckled ping-pong ball is invariant only under rotations through any angle about that axis, and reflections in planes that contain that axis. These form another group. If you think about what these symmetries do to the plane perpendicular to the axis, you'll see that this is essentially just O(2), the orthogonal group in two dimensions.

What is the relation between these two groups? We keep saying that the spherical symmetry O(3) is 'broken' to yield the circular symmetry O(2). What do we mean? We mean that some of the symmetries of the sphere are removed from consideration: they are no longer symmetries of the buckled sphere. On the other hand, *all* of the 'circular' symmetries of the buckled sphere are also valid symmetries of the original unbuckled sphere. In short, O(2) contains fewer symmetries than O(3) – not in the numerical sense, for both groups contain infinitely many transformations, but in the sense that O(2) is a *part* of O(3).

If one group is a part of another group, then we say that the smaller one is a *subgroup*. When symmetry breaks, the symmetry of the resulting state of the system is a subgroup of the symmetry group of the whole system. So symmetry-breaking is a change in the symmetry group, from a larger one to a smaller one, from the whole to a part.

For example, Figure 2.12 shows a famous physics experiment, the Chladni plate. This particular picture was published in 1834: it's taken from *Of the Connexion of the Physical Sciences* by Mary Somerville, one of the great women mathematicians of the ninteenth century, and an accomplished popularizer of science. The apparatus is a square plate, supported at its centre, and sand is sprinkled on it. The bow of a violin makes the plate vibrate, and the sand collects along the *nodal lines*, where the plate is stationary. The apparatus has square symmetry, and so do some of the pictures – namely numbers 4, 5, 7, 9, 10, 11, 13, and 14 counting from the top and along rows. In the rest, square symmetry is broken: they have fewer symmetry transformations. Several different subgroups of D_4 are visible in the figure, and you can gain some valuable hands-on experience by working out the symmetries of each pattern. Call the four rotations 0, 90, 180, 270, and the four reflections H, V, D, U (horizontal axis, vertical, diagonal sloping down from left to right, diagonal sloping up from left to right). Apply all eight transformations to each pattern, and record those that leave it unchanged. For example, pattern 1 is unchanged by 0, 180, D, and U; pattern 2 by 0, 180, H, and V; and pattern 15 by 0 and U.

The general theory of symmetry-breaking starts from this point of view, and tackles questions like 'Which subgroups can occur?' and 'When does a given subgroup occur?' The answers can be very precise: for example, James Montaldi has used this approach to prove that a tetrahedrally symmetric object can vibrate in either 27, 39, or (rarely) infinitely many different ways (and no other number!), depending on the sign of one particular quantity. The result applies equally well to a rubber tetrahedron, four balls on springs, or a methane molecule – the physical system makes no difference to the numerology, only the symmetry matters. We won't drag you through the technical aspects of the answers to such questions, but they provide the mathematical backbone of the theory of symmetry-breaking, and you'll see some of their consequences as we progress.

Figure 2.12 Chladni patterns, formed by sand on a vibrating plate. The plate has square symmetry, but many patterns have less

3

Where Did It Go?

> The truth as we see it today is this: The laws of nature do not determine uniquely the one world that actually exists.
>
> Hermann Weyl, *Symmetry*

We've now discovered that Curie's Principle isn't as straightforward as it may seem. It's not so much false, as liable to mislead those who fail to appreciate its subtleties. The symmetry of a particular state of a system *can* be less than that of the system as a whole. Unlike energy, symmetry need not be 'conserved'. Nevertheless, when symmetry breaks, we're left with a nagging feeling that it must have *gone* somewhere. It can't just vanish. Can it?

Where Does the Symmetry Go?

Good question! The easiest way to answer it is for you to build (or at least imagine) a 'catastrophe machine' (Figure 3.1), and experiment with it. This was invented by Christopher Zeeman in 1969, for a rather different purpose: to illustrate 'Catastrophe Theory', a theory of sudden changes in slowly varying systems. We're going to use it to show that symmetry is not so much *broken* as *spread around*. By following the instructions we're about to give you, you can make a workable model; though if you want something more permanent you have to do a more thorough engineering job. (*Warning*: the machine works best if there's a fair amount of friction at the pivot. The Engineering Department at an Australian University once made one with a ball-race for a pivot, and it spent all its time oscillating to and fro, instead of settling down as required.)

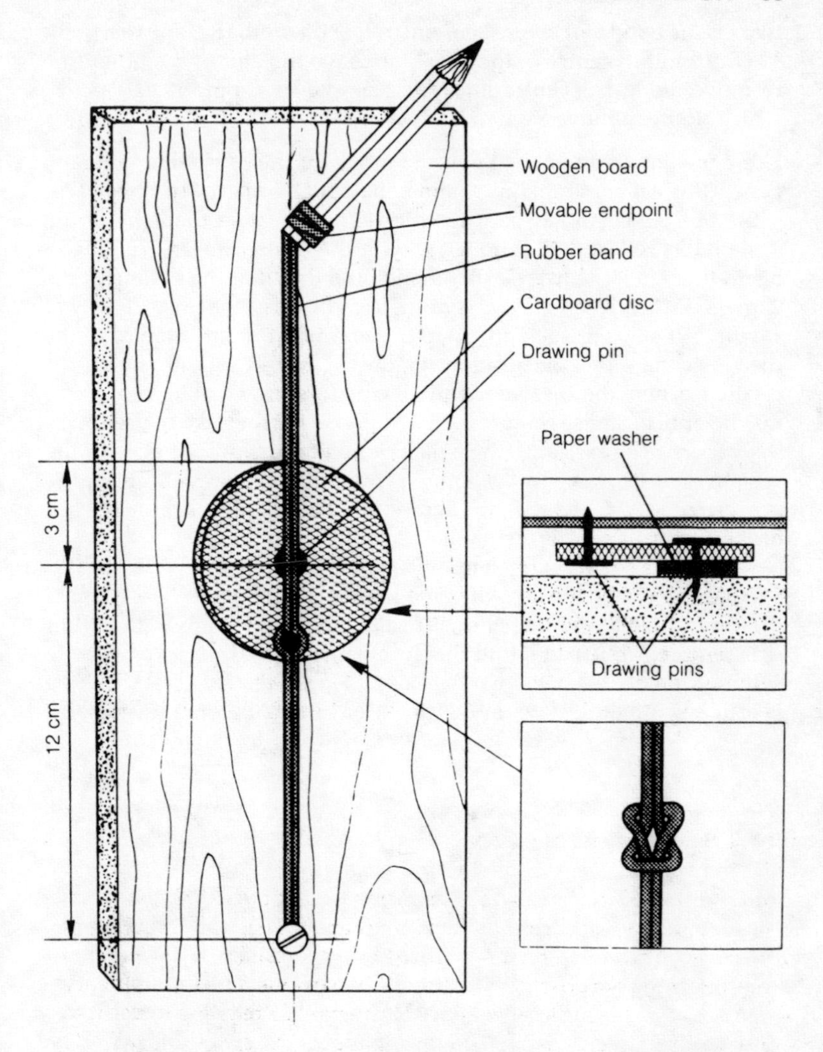

Figure 3.1 Zeeman's catastrophe machine; an experiment in symmetry-breaking

Catastrophe machine instructions

Attach a circular disc of thick card, of radius 3 cm, to a board, using a drawing-pin (thumbtack) and a paper washer. Fix another drawing-pin near the rim of the disc with its point *upwards*. To this pin attach two elastic bands, of about 6 cm unstretched length. Fix one to a point 12 cm from the centre of the disc, and leave the end of the other free to move along the centre line, for example by taping it to a pencil which you can move by hand.

Zeeman allowed the pencil point to move off the centre-line, thereby generating interesting jumps in behaviour, or 'catastrophes'. We want the pencil to stay on the centre-line because then the entire system has reflectional symmetry, with the centre-line as axis. If you begin to stretch the free elastic, then you'll find that the system obeys Curie's Principle and stays symmetric; that is, the disc does not rotate, but stays at such an angle that the point of attachment to the disc is at the 6 o'clock position (Figure 3.2a). But as you stretch the elastic further, the disc suddenly begins to turn – maybe clockwise, maybe anticlockwise (Figure 3.2b). Suppose for the sake of argument that it turns clockwise, to the 5 o'clock position. Now the state of the system *fails* to have reflectional symmetry: if you reflect in the symmetry axis, 5 o'clock flips across to 7 o'clock. The symmetry has broken, and Curie's Principle has failed.

Where has the missing symmetry gone?

Hold the elastic steady with one hand and use the other to rotate the disc to 7 o'clock, the symmetrically placed position on the other side (Figure 3.2c). You'll find that it remains there. Instead of a single symmetric state we have two *symmetrically related* states. In fact, there is also a symmetric state, at 6 o'clock, but it has become unstable.

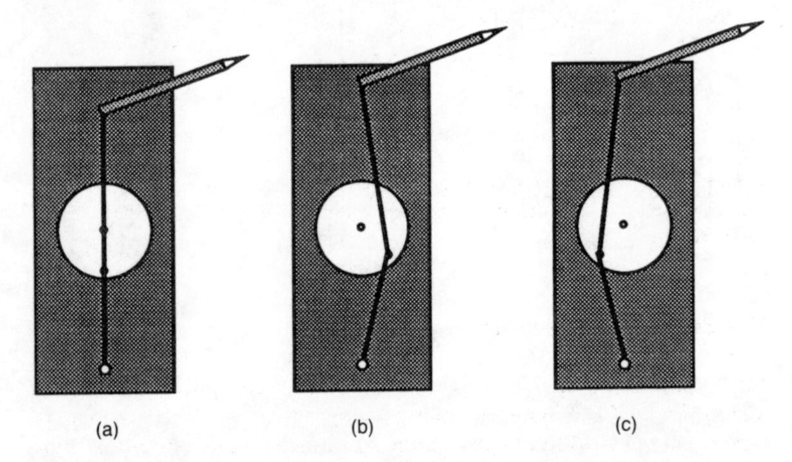

Figure 3.2 Various states of the catastrophe machine. (a) Symmetric state. (b) Perturbed state. (c) Reflected state

We can summarise this behaviour in a *bifurcation diagram*, which graphs the possible steady states of the system (represented by the position of the point at which the elastic is attached to the disc) against the extent to which the free end is stretched. The graph in this case is shaped like a pitchfork (Figure 3.3), and we're seeing an example of *pitchfork bifurcation*, a typical phenomenon in systems with a reflectional symmetry. It's really a 'trifurcation' rather than a bifurcation; but if we ignore unstable states, 'bifurcation' is a reasonable name. Indeed the name is now used to describe any qualitative changes in the states of a system, and not just splitting into two.

The pitchfork shape is symmetric: if you reflect it in the horizontal axis, it looks exactly the same. We'll now investigate why this happens.

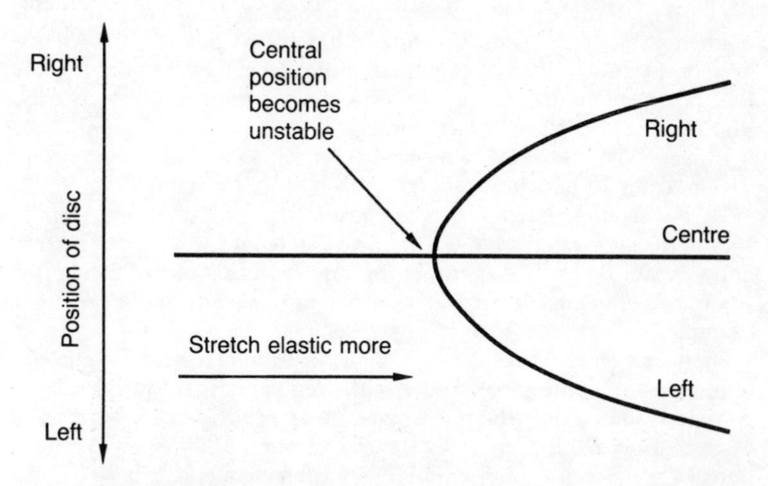

Figure 3.3 Pitchfork bifurcation diagram shows how the position in which the catastrophe machine sits depends upon the amount by which the elastic has been stretched

The Extended Curie Principle

In a general symmetric system, with a more complicated group of symmetries, there may be many symmetrically related states, of which only one occurs in any given experiment. For example, let's go back to the buckling ping-pong ball, which we've seen breaks symmetry from spherical to circular, forming a dent that has rotational symmetry about some axis. Consider only the ideal, perfect sphere. Where can the axis be? We've seen the answer to that, too: just like the 600-pound gorilla in the joke book, it can be anywhere it wants to be. However, the shape of the buckled ping-pong ball is always the same, for a fixed compressive force, wherever the axis is. To get from one possible buckled state to any other, you just have to rotate the sphere until one axis moves to where the other one is. And that rotation is one of the symmetries of the original, perfect system.

This behaviour is confirmed by a simple mathematical analysis of the equations for any symmetric system. Suppose you have such a system, and you select some state that it can take up, (breaking the symmetry or not). Now apply any of the symmetry transformations of the system to that state. Curie's Principle, in its simple-minded form, says that the result should be *the same* state of the system, but as we've repeatedly emphasized, that's not always correct. What the mathematics *does* tell us is that the result is *another* possible state of the system. (Actually, it *can* sometimes be the same state, but only when the transformation you apply is one of the 'unbroken' symmetries that remain.) Let's give this a name.

Extended Curie Principle: physically realizable states of a symmetric system come in bunches, related to each other by symmetry.

To put it another way, a symmetric cause produces one from a symmetrically related *set* of effects. The Extended Curie Principle isn't quite as simple or as elegant as the original version, but it has the advantage of being correct. Let's check it out on some of our previous examples.

First, remember those drops of dew formed on an idealized, infinitely long, thread of spider silk – regularly spaced and nearly spherical. They don't share the symmetries of the thread, which are *all* rigid motions of a line: translations through any distance, reflections in any point. Their symmetries form a smaller list: translations through integer multiples of the spacing between droplets, and reflection in any point that either lies at the centre of a droplet, or halfway between two of them.

According to the Extended Curie Principle, if we take such an arrangement of droplets, and apply any symmetry of the line, that is, any rigid motion whatsoever, we obtain another possible arrangement of droplets. For example, suppose we translate the droplets through an arbitrary distance. The result is just to slide them all along: we get a similar-looking spiderweb, complete with regularly spaced dewdrops, but they're in *different places*. In other words, given that regularly spaced droplets can form at all, they can form starting from any point whatsoever of the thread. The distributions of droplets so obtained all look the same – but they're in different places. That makes remarkably good sense, because on an infinite thread, every point looks exactly the same as every other point: why choose a particular one?

What about the reflections of the line? Because the individual droplets have left-right symmetry, flipping the line of droplets end for end produces the same result as sliding it along. We just pick up the same set of 'new' states again. To see why, draw a picture, and look at it in a mirror. It wouldn't work for a line of left boots, which would become a line of right boots when reflected; but a line of bilaterally symmetric objects, say people, still looks like a line of people when viewed in the mirror.

For argument's sake, suppose that there's a state of the system in which the droplets are regularly spaced, but not left-right symmetric. Perhaps they bulge to the left. Then the Extended Curie Principle would tell us that, by applying a reflection, we can deduce that there must also exist a state in which the droplets all bulge to the right. See: not *the same* solution, but one related to it by the symmetry. Symmetry is shared, not broken.

Next, the drop of milk. As we develop our argument, you'll recognize that we've already anticipated it to some extent, when discussing that paradoxical fact that the circular symmetry of the drop is broken when it splashes. With the particular conditions that prevailed in the experiment recorded at the front of *On Growth and Form*, the possible states are 24-pointed crowns with 24-fold rotational symmetry (and also, like the starfish, a set of reflectional symmetries, this time 24 of them). The original system is invariant under all possible rotations and reflections that fix the line down which the centre of the drop falls; the resulting state is invariant only under 24 rotations and 24 reflections. The symmetry breaks from O(2) to D_{24}. What do we get if we apply an arbitrary rotation to a crown-shaped drop? We get another drop of identical shape, but with the spikes of the crown in a different place. That's the only difference. If we reflect the crown-shaped drop in an arbitrary plane through its centre, we get another of the same shape, but rotated to some other position.

For a third and final example, consider Bénard convection. In an idealized model this takes place in an infinite, uniform plane layer of fluid. The symmetries of the system are all possible rigid motions of the plane – all translations, all rotations, all reflections, a huge symmetry group. Experiment, and various calculations, show that one way for the system to break symmetry is by forming a hexagonal lattice – a honeycomb – of convection cells. These are invariant under a subgroup, consisting of *some* translations, *some* rotations, and *some* reflections – the rotations and reflections that leave a hexagonal cell invariant, and the translations between one cell and another.

What do we get when we apply an arbitrary symmetry of the whole system, an arbitrary rigid motion, to a honeycomb of cells? We get another honeycomb, but one that is in a different position or has a different orientation. Again this makes sense: in a perfectly uniform infinite plane, every point is as good as any other, and every direction as good as any other. The physical system has to choose where to form a cell and in what direction to lay out its sides; but in the mathematical idealisation, that choice is an arbitrary one. Again solutions come in symmetrically related bunches, and the Extended Curie Principle is valid.

So now the dreadful truth is out: symmetries are not so much broken as shared around. The phrase should really be symmetry-*sharing*, not symmetry-breaking. Despite this, we continue to talk of *broken* symmetry, because that's the conventional terminology of our subject. It's a reasonable phrase to use, because in experiments you usually can observe only one member of the symmetrically related bunch of solutions that the mathematics guarantees. A buckling sphere can't buckle into two shapes at the same time. So, while the full potentiality of possible states retains complete symmetry, what we observe seems to break it. A coin has two symmetrically related sides, but when you toss it it has to end up either heads or tails: not both. Flipping the coin breaks its flip symmetry: the actual breaks the symmetry of the potential.

There's a related use of the phrase 'broken symmetry', which we mention to avoid confusion if you read other sources, and because we occasionally need to refer to it and distinguish it from our current usage. The type of symmetry-breaking that we've just been talking about occurs even though the overall system continues to possess perfect symmetry. It's just the solutions that break, lose, or if you prefer share, their symmetry. We call this *spontaneous* symmetry-breaking to show that it's a natural consequence of internal instabilities, rather than being imposed from outside. But there's a second type, *induced* symmetry-breaking, that *is* imposed from outside. It occurs when some external agent changes the system in a way that destroys its symmetry.

A ping-pong ball buckling under a spherically symmetric force undergoes spontaneous symmetry-breaking. But we could imagine applying a force that's not spherically symmetric at all, or altering the shape of the ping-pong ball to an ellipsoid, or a featureless lump. It must still do *something* when it buckles, but now we'd be surprised if what it did were spherically symmetric. That's induced symmetry-breaking. It's less paradoxical than spontaneous symmetry-breaking, because we don't expect the symmetry to be preserved in any case. It obeys Curie's Principle instead of (apparently) violating it. However, it has important applications, because the behaviour of a system that is in some sense 'close' to a perfectly symmetric one often retains some traces of that symmetry. Induced symmetry-breaking is a method for handling approximate symmetries. We don't want to get sidetracked, so unless we say otherwise, when we use the phrase 'symmetry-breaking' we'll mean the spontaneous kind.

Time Symmetries

Symmetries can occur in time as well as space. For example, ignoring small perturbations caused by the other planets, the Earth goes round the Sun in one year – meaning that at intervals of exactly a year, the Earth returns to the same position relative to the Sun. This is true not just of its starting point, but of every point in its orbit. If you wait in space until the Earth comes by, then you know that you'll have to wait exactly a year for the next such occasion. Unlike city buses, where you wait random times and then three arrive at once, the Earth's orbital motion is *periodic* – it repeats at regular intervals. The size of such an interval is called the *period*: for the Earth's orbital motion round the Sun, the period is one year.

Periodic behaviour has its own kind of symmetry – as you'll see immediately if you consider the effect of the transformation 'wait one year'. The Earth will then be in exactly the same position relative to the Sun as it was before, and that's what 'symmetry' means: after the transformation, the system appears unchanged. However, unlike the symmetries we've been looking at so far, this kind of symmetry involves changes in time rather than in space. Indeed 'wait a year' is a *translation* of time, a fact that is most readily apparent if in traditional fashion we represent time by a one-dimensional line (Figure 3.4). Every point on the line – every instant of time – slides along through an interval of one year.

Of course, once waiting a single year gives a symmetry of the Earth–Sun system, then waiting two years or three years or any number of years in the future will also give a time symmetry. Time symmetries are like that – once you have one, you get a whole slew of them. This may be why they made so many sequels (if that's the right word) to *Back to the Future*.

Indeed, there's still more. Once you know that the Earth–Sun system will exactly repeat its dance each year in the future, you *know*

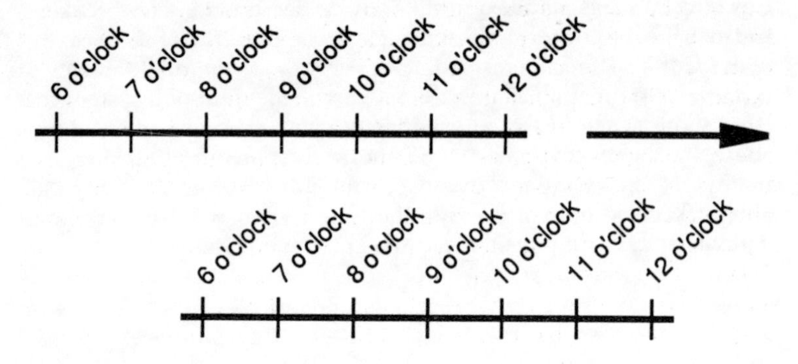

Figure 3.4 Translation along the time axis

that it must have repeated that dance every year in the past. That is, translating backwards in time by one year or two years or any number of years is also a time symmetry.

The realization that symmetries can exist in time opens up a whole new range of enquires. Any system whose behaviour depends upon time – known as a *dynamical system* – may possess temporal symmetries as well as spatial ones. Moreover, since time can be represented by a line, the natural temporal symmetries are the 'rigid motions' of that line. We've already determined what those are: they're all possible translations and reflections. Periodic motion happens to be motion that is invariant under a particular time translation; but what other possibilities are there?

First, we should sort out what the *most* symmetric kind of motion, in the temporal sense, is. After all, we're interested in symmetry-breaking, so the more symmetry we start with, the more possibilities there are for it to break. The most time-symmetric motion must be invariant under all translations of time (and all reflections, for that matter). What does that mean? It means that the system must appear to be exactly the same, whenever you decide to look at it. The system is in an identical state at all instants of time. (Such a system is invariant under all time reflections too, so considering reflections adds no further information.)

'But that means it isn't doing anything!'

Right! Systems that possess all possible time symmetries are those that remain in *steady* states – doing nothing. The archetypal practitioner of steady-state activity is the rock. Geologists, of course, think of rocks as rather more dynamic things; we're talking about behaviour on a human timescale. Admittedly, when you drive down a road and observe that the rocks alongside it look the same as they did yesterday, last week, or even last February, you don't normally exclaim 'What magnificent temporal symmetry those rocks possess!' Once more we've run into the peculiarity of the human mind, that it doesn't recognize extremes of symmetry as constituting pattern. We are far more impressed by the degree of pattern in a periodic phenomenon than we are in one that simply stays put; we tend not to notice that the one that stays put has even *more* pattern.

Time translations, we've already seen, correspond to periodic motion. What about time reflections? Reflecting a line reverses its negative and positive directions; reflecting time interchanges past and future. A motion that is invariant under a time reflection will look the same if you make a film of it and then run the film backwards. The point about which it is reflected – the 'time mirror' – will then be the unique instant of time when the real world is identical to that in the reversed film.

In other words, the motion after the mirror-point is the same as the motion before, but runs backwards. Do such bizarre motions exist? They do; in fact they're quite common. For example, suppose you throw a ball vertically upwards, and that it reaches the top of its flight at precisely noon. The motion before noon is upwards, slowing to an instantaneous stop at noon precisely; the motion after noon is downwards, starting from rest. That is, the afternoon motion is the reverse of the morning motion. So the ball is a system that has a reflectional time symmetry. We say that it is *time-reversible*. Another example is a pool ball hitting a cushion at an exact right angle: the motion after the instant of impact is the reverse of the motion beforehand.

Usually, if you throw a ball into the air, it doesn't travel exactly vertically. It might, for example, move away from you towards the north. The time-reversed motion will then be toward the south, which is no longer the same. However, the ball's motion still has a symmetry that involves reversing time, but this must be combined with a spatial reflection that interchanges north and south. In the ideal case (motion in a vacuum) the path of the ball is a parabola; and the spatial mirror plane is a vertical one, at right angles to the direction of motion of the ball, passing through the highest point of its flight. A pool ball that hits a cushion at an angle has a similar kind of symmetry: reverse time and reflect in a plane at right angles to the table and to the cushion, passing through the point of impact.

This type of time-reversal symmetry – time reflection *plus* some spatial transformation – is exceedingly common. If you make a film of a cat running, reverse it, *and* reflect the cat from front to back, then the pattern of footfalls is almost identical to that of the original cat (Figure 3.5). In fact, time-reversal symmetry is a universal phenomenon, in the following sense. The motion of matter can be described mathematically, and – except when relativistic or quantum effects are important – it is customary to use Newton's Laws of Motion. These prescribe how matter will move under the effect of forces. Now, a simple mathematical consequence of the form of Newton's Laws is that all physical motions have a symmetry, combining time-reflection with a 'spatial' symmetry – though in a space somewhat larger than

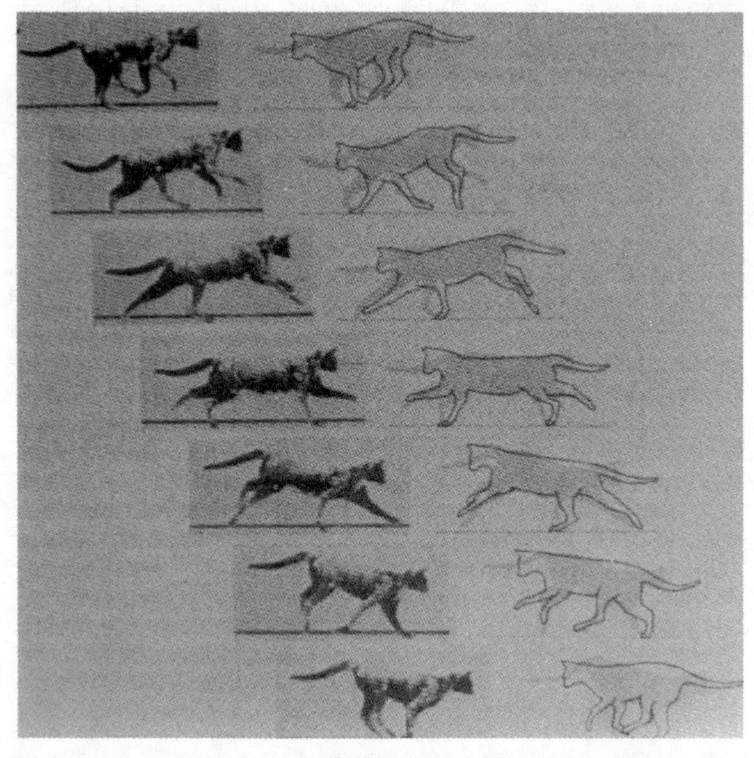

Figure 3.5 Time-reversible pattern of cat footfalls. Left: a galloping cat. Right: the same sequence (shaded), upon which a time reversed-sequence (outline) is superimposed in a way that fixes the cat's centre of gravity. The reversed cat's paws match the original almost exactly. Notice that the cat, as well as time, must be reflected

the one we live in. Think of a system of bodies (anything from atoms to galaxies) moving in a manner that obeys Newton's Laws. If at some instant you simultaneously reverse the direction of time, and the velocities of every component body – a kind of collective 'bounce' off an imaginary mathematical wall – then the motion will continue to satisfy Newton's Laws. Reversal of velocity is not something that you can perform with an ordinary mirror; but mathematicians studying Newtonian dynamics habitually use an extended notion of 'space' in which the usual three coordinates are supplemented by three more velocity coordinates, and an appropriately generalized 'mirror' can reverse all velocities while leaving all spatial positions unchanged.

It is this 'universal' time-reversibility of motion that lets TV producers play games by running videotapes backwards. One episode of *Red Dwarf,* a science fiction sitcom popular in the UK, involved time-reversals of people drinking beer, eating an eclair, dropping sugar lumps into tea, and having a fight in a bar. Drinking and eating have a midly repulsive appearance, to be sure; but such effects have a disturbing aura of plausibility. As the sugar-lumps pop up out of the teacup and land on the spoon, they move *just like normal sugar lumps*. As bodies leap upright from the floor and deliver devastating blows to opponents' hands with their chins, as broken mirrors reassemble themselves and expel wine bottles, which speed towards hands waiting to catch them, the mathematical dynamics looks impeccable – it's only the social dynamics that are weird.

The universal time-reversibility of Newtonian mechanics poses a deep philosophical question: *why does time appear to flow in one definite direction?* Innumerable learned and popular treatises hinge upon this topic, with explanations ranging from chaotic dynamics to quantum theory. A notable instance is Stephen Hawking's recent bestseller *A Brief History of Time*. It's a vast mystery, leading directly into huge areas of human ignorance. The explanation might equally well be related to alternatives to Newton's Laws such as quantum mechanics, to the way we currently interpret those laws, or to the physical nature of human consciousness. Those are issues too deep for us to discuss in this book; but it's striking that a simple symmetry property can plunge us headlong into such a philosophical morass as the problem of the nature of time.

Wobbles

Having established the basic types of temporal symmetry, we're now ready to discuss how it can break. Where do periodic motions come from? Various mathematicians, among them Henri Poincaré, Aleksandr Andronov, and Eberhard Hopf, discovered a fundamental dynamical process usually known as *Hopf bifurcation*. This creates periodic motion from a steady state.

You can experiment with Hopf bifurcation if you possess a record-player with continuously variable speed. If not, treat this as a thought experiment. Place a pudding-basin on top of the turntable, accurately centred, and put a ping-pong ball inside it. The basin should be curved everywhere, and *not* have a flat base; and it should be circularly symmetric. Look down from above, and observe how the position of the *centre* of the ping-pong ball varies when the turntable is rotated at different speeds. At zero speed, it sits at the bottom of the bowl, in the middle of the turntable: this is a steady state. It *continues* to sit there if the turntable revolves *slowly*: the steady state persists. (The ping-pong ball rotates, but we are supposed to be looking only at its centre, which remains fixed.) At a suitably higher speed, however, the ping-pong ball climbs the side of the bowl, pushed by centrifugal forces. From out overhead viewpoint its centre is now going round and round, a *periodic* motion. In between there is some critical speed at which the ball leaves the steady state at the centre and begins to climb up the side of the bowl: from that instant onwards its motion has become periodic.

The simplest way to describe Hopf bifurcation is as the onset of a wobble. The idea is that the system is influenced by some external variable as well as undergoing its own internal dynamics. At first the system is in a steady state, and does nothing; but as the external variable changes, a very slight wobble develops, which then grows until it becomes pronounced. In our ping-pong ball example the 'wobble' is the circular motion that develops as it climbs the bowl – we're not suggesting that the ball wobbles from side to side.

While the system is in a steady state, its temporal symmetries are all translations and reflections of time. As soon as the periodic wobble appears, the temporal symmetry breaks, and now the only temporal symmetries are translations through whole number multiples of the period. So steady states turn into periodic wobbles through a process of temporal symmetry-breaking.

Wobbles – more properly called periodic oscillations – have two important numerical features. One is the *amplitude*, or size, of the oscillation. In the ping-pong ball example the amplitude is the radius of the circle around which it moves. The other is the *phase*; for the ping-pong ball this is its angular distance round the circle at some chosen instant of time. In general the phase of an oscillation is determined only relative to another oscillation, of the same general form but translated in time. The phase is that time difference, measured as a fraction of a period. For example, if you watch the Earth going round the Sun, and on 1 April you start off a fake planet Earth where the real one was three months earlier on 1 January, then the phase difference between them is one-quarter of the period, because three months is one-quarter of a year. You need two oscillations because 'difference' makes no sense for a single object: there's got to be another to differ *from*. The entire process of Hopf bifurcation has *phase shift symmetry*, meaning that if any particular oscillation solves the mathematical equations, so does any phase-shifted (time translated) version of it. The fake Earth will also obey Newton's Laws of Motion. Another way to say this is that nature doesn't choose a particular time origin. Dates are human conventions, and – all else being equal – the result of an experiment doesn't depend upon the day of the week that you perform it on.

The Hosepipe

Rather than give an example of Hopf bifurcation in its own right, we'll take matters one stage further, and examine the combination of spatial and temporal symmetries that governs the onset of wobbles in systems that start with a certain amount of spatial symmetry. You've already been introduced to two possible experiments on symmetry-breaking in your own home – the catastrophe machine and the frying-pan. Here's another, which leads us into a rather different, and very important, range of phenomena: *symmetric Hopf bifurcation*.

An accessible, but messy, example of symmetric Hopf bifurcation occurs when a hose of circular cross-section is suspended vertically, nozzle downwards, with water flowing steadily through it. Here the internal dynamical variables are the positions and velocities of points of the hose, and the external variable that 'switches on' the wobble is the speed with which the water flows. You can experiment using flexible rubber tubing of the kind found in chemistry laboratories, about 5 mm in diameter, although a thought experiment is drier. Because the hose has circular cross-section, the system is circularly symmetric about an axis than runs vertically up the centre of the hose. That is, it is invariant under all rotations about that axis, and under all reflections in planes passing through that axis. And indeed, if the speed of the water is slow enough, the hose just remains in this vertical position, and the system retains its circular symmetry.

However, if you turn the tap on further, then the hose begins to wobble. In fact there are two distinct kind of wobble, and which one

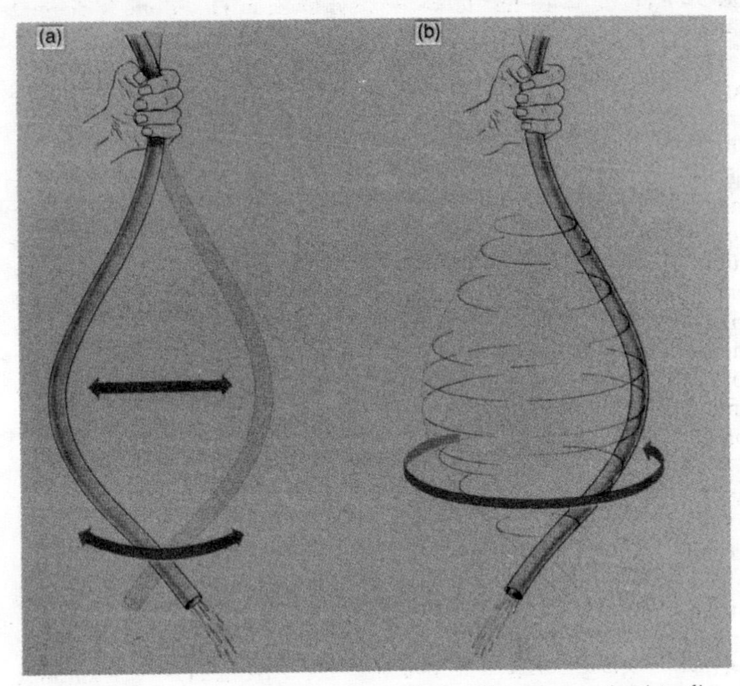

Figure 3.6 Oscillations of a hosepipe. (a) Pendulum-like planar motion (standing wave). (b) Circular motion (travelling or rotating wave)

occurs in your experiment will depend on the length and flexibility of the tubing. In one kind, the tube swings from side to side like a pendulum (Figure 3.6a). In the other, it goes round and round, spraying water in a spiral (Figure 3.6b). Similar effects are often observed when children wash the family car and 'accidentally' let go of the hose. When the hose starts to wobble, it no longer possesses circular symmetry about a vertical axis: indeed the symmetry breaks in two distinct ways, and the total temporal symmetry of the steady state breaks as well.

The perfectly symmetric state of the system is to hang vertically. But, when the speed with which the water is flowing becomes large enough, that state becomes *unstable*. The symmetric (vertical) state still exists mathematically, but you don't observe it in practice because any tiny random deviation tends to grow. Since the symmetric state can't occur, then naturally the system has to do something else, which perforce has to be less symmetric. However, it doesn't become *totally* asymmetric.

We've said that one possibility is to swing to and fro in a plane. If you think of that plane as a mirror, then the swing is symmetric under reflection in the mirror. That's an example of a *standing wave*. The other type of periodic oscillation that can occur is to swing in circles. You might think that this motion has circular symmetry, but that's not true. If you rotate the system through some angle, then it doesn't look *exactly* the same. It's the same general kind of motion, but in a different place at a given time. You have to rotate the system, *and translate time*, before you reproduce the identical motion. If you rotate the system, and apply a suitable time delay, then it looks *exactly* the same as before. And in this case the time delay is the same as the rotation, in the sense that a rotation of some faction of a turn needs a time delay of the same fraction of a period. Such a motion is called a *rotating wave*.

So in Hopf bifurcation with circular symmetry, when the perfectly symmetric state becomes unstable, the symmetry can break either to a standing wave or to a rotating wave. The standing wave has a purely spatial symmetry – reflection in its plane. The rotating wave has a mixed spatio-temporal symmetry: rotate through an angle and translate time by a corresponding amount.

However, in accordance with the Extended Curie Principle, the circular symmetry hasn't *totally* vanished. Consider the standing wave, symmetric in a mirror plane. Choose any vertical plane passing through the symmetry axis. Then the hose can oscillate to and fro in that plane. How are all those planes related? They're all rotations of each other. Instead of there being a single state of the system, unchanged by all rotations – that is, a fully symmetric state – you get *lots* of less symmetric states, all related to *each other* by rotations. The whole set of motions still has circular symmetry, in the sense that if you rotate any possible motion, you get another one in that set. The same goes for reflections, though because standing waves already have reflectional symmetry, reflections don't produce anything new.

Similarly, if you take a rotating wave motion and rotate or reflect it, you get another possible rotating wave. The rotations have very little effect: they act just like time translations. For this type of oscillation it's the reflections that count: reflections turn clockwise waves into anticlockwise ones.

In a given system, at most one of these types of oscillation is stable when it first switches on. Possibly neither are. The point is that they can't both be stable at the same time. The system *chooses* one or the other – or neither. This *selection rule*, that a choice between these two possibilities must be made, is a model-independent phenomenon – but the choice actually made depends on the model.

Coupled Oscillators

In a few chapters' time we'll need to refer to similar results for a slightly more esoteric case, Hopf bifurcation with dihedral group symmetry. You'll recall that the dihedral group D_n is the group of symmetries of a regular n-sided polygon. At first sight it's not so obvious where such symmetries might naturally arise. Hopf bifurcation with D_5 symmetry would no doubt be ideal for describing a wobbling starfish, but there's not much call for that kind of analysis. In fact, however, there's an important setting in which precisely this type of symmetry arises: the theory of coupled oscillators.

An *oscillator* is anything that can oscillate, that is, wobble periodically. Hopf bifurcation is one of the most basic and most common mechanisms by which an oscillator can 'switch on'. There are others, but we don't need to describe them here. Symmetries arise naturally when a system of identical oscillators is linked together in a network. Such a system is said to be *coupled*, the word 'coupling' being used to describe the influence of one oscillator upon another. For example, the pendulum in a grandfather clock is an oscillator. If you stand two grandfather clocks next to each other on a floorboard that's badly nailed down, so that it's sloppy and can vibrate, then the oscillations of the pendulum in one clock are transmitted via the floorboard to the casing of the other clock, and thence to its pendulum. The usual result is that the two pendulums *entrain* each other – begin to oscillate synchronously. For this reason, watchmakers learn not to hang several watches on the same rail when they're trying to regulate their timing. We, however, wish to understand what the watchmakers wish to avoid: the effect of coupling between several oscillators.

The simplest case is when two identical oscillators are symmetrically coupled. Think of two watches of the same make and style hung on a single rail. This system has one nontrivial symmetry: swap the two oscillators. The symmetry group is the cyclic group Z_2. Now, the theory of Hopf bifurcation with Z_2 symmetry tells us that there are two different types of Hopf bifurcation in such a system, so exactly two types of oscillation pattern are possible. In the first, the in-phase pattern, both oscillators behave identically. This motion doesn't break the Z_2 symmetry. The second pattern does break it: this is the *out-of-phase* pattern. One oscillator lags behind the other by exactly half a period. Two watches that are going 'tick-tock' in phase will tick simultaneously, tock simultaneously. If they are out of phase, then when one ticks, the other tocks, and vice versa.

Dihedral group symmetry occurs when several identical oscillators are connected in a ring, as shown schematically in (Figure 3.7). The polygonal symmetry is evident to the eye. The mathematics of such a system is quite intricate, but we can convey the flavour of it by describing just one representative case: three oscillators. The symmetry group is then D_3, that of an equilateral triangle: three rotations, through 0°, 120°, and 240°, and three reflections.

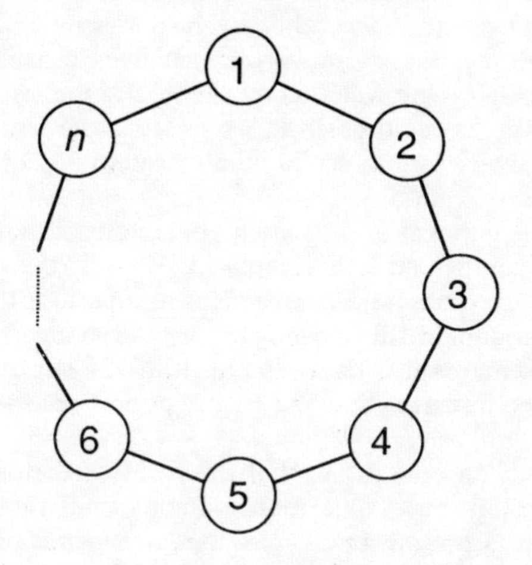

Figure 3.7 A ring of n *identical symmetrically coupled oscillators*

For three coupled oscillators, as for two, there are two distinct kinds of Hopf bifurcation. The first is analogous to the in-phase solution for two oscillators: the symmetry is unbroken, and all three behave identically. Three watches, all going 'tick' together and 'tock' together. (Watches of a feather tock together).

The alternative, however, is far stranger. Instead of a single out-of-phase oscillation, three distinct patterns become possible simultaneously. However, at most one of them is stable: just which one depends upon the detailed dynamics and the nature and strength of the coupling. The three patterns are:

- *Discrete rotating wave* All three oscillators perform the same motions, but successive oscillators round the ring lag behind each other by one third of a period. Think of three watches, all going 'tick-tack-tock'; but when one goes 'tick' the next goes 'tack' and the third 'tock', chasing each other round the ring like ponies on a merry-go-round. This is a discrete analogue of the rotating wave motion of a hosepipe.

- *Reflectionally symmetric oscillation* Two oscillators (choose any pair) behave identically; the third does something else. Think of three watches: two go 'tick' together and 'tock' together, and the third goes 'thud-bonk' instead.
- *Phase-shifted reflectionally symmetric oscillation* Two oscillators (choose any pair) behave identically, except that they are out of phase with each other by half a period. The third does something else; moreover, it oscillates twice as rapidly as the other two. One watch goes 'tick-tock'; the second 'tock-tick', and the third 'thud-thud', repeating the same basic 'thud' *twice* every period of the whole system.

Each of these patterns has its own characteristic symmetry. In the first, if we cycle the oscillators, replacing each by the next one round the ring, *and* translate time by one-third of a period, then the system remains invariant. In the second, if we interchange two oscillators, the system remains invariant. In the third, if we interchange two oscillators *and* translate time by half a period, the system remains invariant.

The curiosity in case three, that the third oscillator runs twice as fast as the others, is a consequence of this final symmetry; for the oscillator that is not swapped must be *half a period out of phase with itself*. Remarkable! This means that if we translate time by half a period, it is doing the same thing as before. That is, its true period is half that of the other two oscillators, so it wobbles twice as fast.

From the group-theoretic viewpoint, this strange result emerges naturally and inevitably; and it is fully confirmed by computer experiments. It would take remarkable insight – or luck – to guess it any other way. The results for four or more oscillators coupled in a ring are similar, but more intricate, and the greater the number of oscillators, the more possible patterns there are. But they can all be deduced by applying the same group-theoretic principles. Moreover, similar methods can be used to study other networks of oscillators. And coupled oscillators are important in understanding all kinds of natural phenomena, from ocean waves to the vibrations of a methane molecule. Indeed you carry a *very* complicated network of coupled oscillators around with you: your own nervous system. Moreover, as we shall see in chapter 8, the symmetries of at least some of its small subsystems are clearly visible. Pace up and down and think about it.

4

Forever Stones

> It was domination by a beauty so pure that it held a kind of truth, a divine authority before which all other material things turned, like the bit of quartz, to clay. In these few minutes Bond understood the myth of diamonds, and he knew that he would never forget what he had suddenly seen inside the heart of this stone.
>
> Ian Fleming, *Diamonds are Forever*

When the Stewart family takes a trip to London, one almost inviolable ritual is the visit to the Geological Museum. And one of the favourite exhibits in the museum is its collection of crystals. It's a fascination that we share with a large proportion of the human race. Excavations in the Lower Cave at Chou-k'ou-tien show that Peking Man – *Homo Erectus* – collected quartz crystals, between 250,000 and 400,000 years ago. In 3000 BC the Mesopotamians and Egyptians engaged in extensive trade in lapis lazuli: we know this because the stones have been found in Egyptian bogs, but the only large source is in Afghanistan.

Crystals have a strange and enticing beauty. They sparkle, they glisten with brilliant colours – but above all, they're interesting *shapes*. If anything in the universe looks geometric, it's surely a crystal (Figure 4.1). Salt crystals, for instance, are cubes: you can't get much more geometric than that! Nearly every textbook of crystallography begins by telling its readers that the regular forms of crystals 'obviously' suggest that they are built up from identical units, arranged in a regular fashion. This appears to be a severe case of twenty-twenty hindsight – wisdom after the event. In the development of crystallography not only was that statement unobvious: for centuries it was highly controversial, and for excellent reasons.

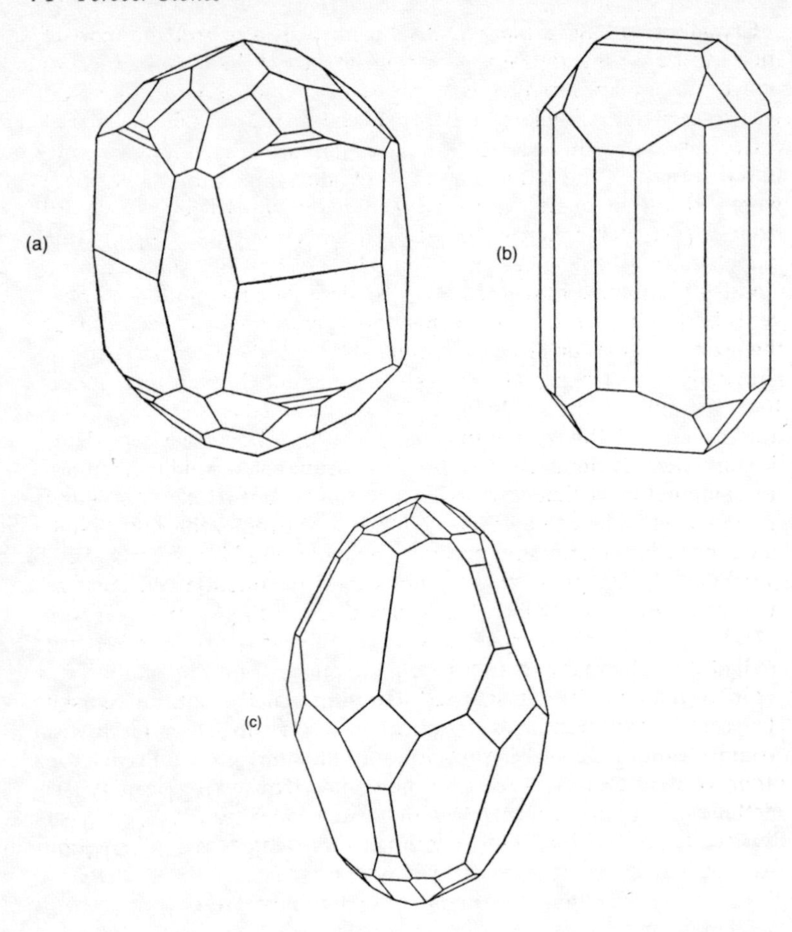

Figure 4.1 Perfect crystals of (a) mercurous chloride, (b) lead sulphate, (c) calcite

Humans value crystals above almost anything else. As our chapter title and quotation allude, they give them to their mates to symbolize eternity. Rubies, emeralds, sapphires – the very names set one's palms twitching. There's no lure like the lure of gemstones. Men have been murdered, even wars waged, over a single crystal. Thieves have risked their freedom and their lives to secure an especially spectacular specimen. For centuries the royal families of Europe have staggered under the weight of their crown jewels.

Crystals also have a mathematical beauty. That's probably part of their intrigue: they appeal to that magical part of the human mind in which mathematics merges with mysticism. Plato thought that God was a geometer; jewellers probably think She's a crystallographer.

Crystals are earthly objects – in fact, Plato thought they were composed entirely of the element earth. (In Plato's time, remember, there were just four elements: earth, air, fire, and water.) They're a far cry from the crystal spheres that the ancients imagined convey the planets across the heavens. Yet, ironically, the first real success of the God-is-a-Geometer movement was to describe the motion of heavenly bodies in mathematical terms. Johannes Kepler's discovery that the orbit of Mars is an ellipse was published in 1609. Newton's Law of Gravitation, which explains the ellipse and much else, dates from 1687. In contrast, the mathematical laws of crystals went unrevealed until the start of the ninteenth century. At first sight this is puzzling, because the mathematics of crystals is arguably simpler and more fundamental than that of celestial mechanics; but it often takes the human mind longer to perceive simplicity. The regularities of crystalline form, moreover, were for a long time obscured by crystals' great variability, by frequent imperfections, and by phenomena such as twinning, whereby two crystals effectively interpenetrate (Figure 4.2). When Kepler was trying to work out the shape of Mars's orbit, he had a much cleaner set of data to work on.

Or course, you've anticipated the answer: the key to crystal structure is symmetry; and broken symmetry to boot. Why else would the topic appear here? Historically, though, it was as much the other way round: crystals were one of the major sources of the mathematics of symmetry. The history is so convoluted, and so illustrative of the half-blind groping from which major scientific

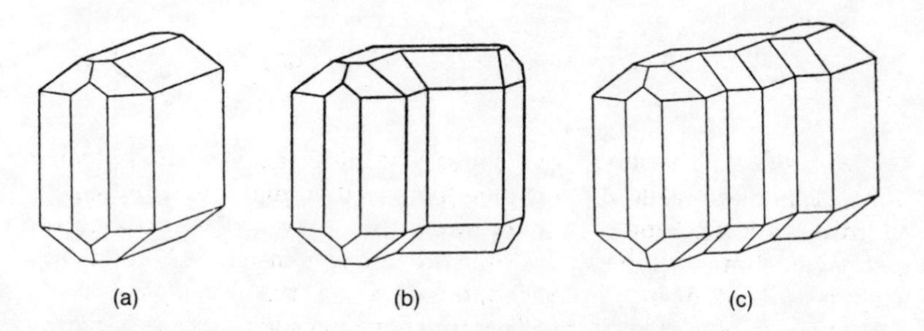

Figure 4.2 (a) Single crystal of potassium sulphate. (b) Twinning of potassium sulphate crystals. (c) Repeated twinning

discoveries can casually emerge – often unappreciated even by those who discover them – that we'll be lenient when allotting space to the historical side of the topic. *That* crystals are symmetric will become apparent along the way. *Why* they are symmetric is a deeper question, which we'll deal with towards the end of the chapter.

Four Greek Theories

The first attempts that we know of, to put crystals on a rational basis, were made, characteristically, by the ancient Greeks. They came up with at least four theories, all interesting because they anticipate aspects of the modern approach. The theories are associated with the Atomists, Plato, Aristotle, and the Stoics.

The Atomists held that every natural object is composed of tiny indivisible particles, which they called atoms. They believed that there are many different types of atom, but that each type occurs in huge numbers. Greek atoms stick together mechanically – hooks and eyes. Lucretius thought that hard substances such as diamonds must be made of atoms that branch and twine together.

Plato's philosophy was syncretic – a polite way of saying that he pinched his ideas from whatever source he could. He took the Pythagorean idea that nature is mathematical, and modified it to make triangles rather than points the basic ingredient. The elements of earth, air, fire, and water, said Plato, are composed of particles shaped like four of the five regular polyhedra. In the *Timaeus* he describes the formation of crystals:

> Of the varieties of earth, that which has been strained through water becomes a stony substance in the following way. When the water mixed with it is broken up in the mixing, it changes into the form of air; and when it has become air, it rushes up towards its own region. Since there was no empty space surrounding it, it gives a thrust to the neighbouring air. This air, being heavy, when it is thrust and poured round the mass of earth, squeezes it hard and thrusts it together into the places from which the new-made air has been rising. Earth thrust together by air so as not to be soluble by water forms stone, the finer being the transparent kind consisting of equal and homogeneous particles.

Aristotle rejected Plato's theories, mainly because he thought that a vacuum can't occur in nature. He claimed that only two regular solids can fill space, the tetrahedron and the cube. (He was wrong about the tetrahedron.) His most interesting idea was that the cohesion of mineral matter is controlled by some internal force.

Finally, the Stoics believed that properties of matter are governed by the *pneuma*, a mixture of fire and air that pervades the universe. The form and properties of matter – for instance, the transparency of a crystal – are a result of the particular mix that occurs there. Various pneumas surge through any material object, pulling it together.

Four theories, all the product of a small amount of observation and a lot of imagination – quaint to the modern mind, yet each containing a germ of truth. The atoms of the Atomists, the geometric form of atoms postulated by Plato, Aristotle's cohesive forces and space-filling solids ... and the Stoics' pneumas, embodying wavelike motion. Of course, by selecting those features that match modern theories, and neglecting those that don't, we can overestimate the cleverness of the Greeks: we're not suggesting they really anticipated quantum theory. But they scored some notable hits, for all their neglect of experiment, and nobody did much better for two millennia.

Snowballs

The puzzle of crystals continued to exercise the thoughts of the learned. Duns Scotus, a thirteenth-century theologian, thought that minerals are alive, that they grow like a plant, and that their regularities of form are akin to those of living creatures – which also exhibit considerable diversity around an 'ideal' form. So did the physician Paracelsus in the sixteenth century. Nicolaus Steno, an anatomist who was also interested in geology, had other ideas. Steno had an experienced observer's 'wise eye', and spotted things that others had missed. Noticing that rocks form natural layers, he deduced that they are formed from deposited minerals. Taking the idea further, he stated that the growth of any body, organic or inanimate, is caused by the accumulation of new particles secreted by a fluid – either an external or an internal one. Plants grow by an internal agency, crystals by an external. A tiny 'seed' – whose creation he frankly admitted he couldn't explain – begins the process; then material accumulates on its facets, attracted by some fluid within the seed, much as a magnet attracts iron filings.

The geometric form of crystals continued to fascinate, and the hexagonal symmetry of snow crystals (Figure 4.3) was a common theme. Johannes Kepler thought about crystals as well as the orbit of Mars and came much closer – very close indeed. In a remarkable essay, *The Six-Cornered Snowflake*, composed as a New Year's present

Figure 4.3 Snow crystals: a variety of designs, all with hexagonal symmetry

for his benefactor Johann Matthäus Wacker von Wackenfels, he suggested that snowflakes must be composed of identical tiny spherical particles arranged in a lattice:

> Granted then that vapour coagulates into globules of a definite size, as soon as it begins to feel the onset of cold. This is reasonable: for the drop is the smallest natural unit of a liquid like water, because, when it is under the size of a drop, its weight does not make it spread more widely ... Second, granted that these balls of vapour have a certain pattern of contact: for instance, square in a plane, cubic in a solid.

But he has problems with the whole idea:

> Now if their material would not admit of being otherwise, our task would be done. But, as was said above, the balls can be arranged in two other ways consonant with their material. Moreover, they can be duly stacked in all three orders, and then shuffled into hybrid patterns.

Kepler knew that there are several regular ways to stack spheres, and he couldn't see good reasons to prefer one over another. Moreover, he was worried about irregularities in the pattern – now known as dislocations – long before anybody knew for certain that there *are* irregularities to worry about. What are the three stacking patterns to which he refers? In the plane there are two regular ways to arrange a lot of identical spheres: in a square lattice, like the pieces on a checkerboard, or in a hexagonal lattice, like the cells of a honeycomb or the red balls at the start of a game of snooker. These arrangements in turn can be stacked in space in several ways. At first sight, there are four combinations; but Kepler observed that they reduce to three (Figure 4.4). 'Square' layers can be stacked so that corresponding spheres are vertically above each other. The centres of the spheres lie at the corners of cubes which stack together to fill space in the obvious way. Alternatively the spheres in each layer can be laid so that they nestle into the gaps between four spheres in the layer below, as greengrocers often do with oranges or apples. With 'hexagonal' layers there are also these two possibilities, aligned or staggered; but staggered layers of hexagonally packed spheres lead to the *same* arrangement as staggered layers of squarely packed ones. The only difference is that the layers of one are tilted in comparison to the corresponding layers of the other. This is not so hard to see: use staggered 'square' packing to build a square pyramid of oranges, and then look at one of its sloping faces: you'll see one layer of a staggered 'hexagonal' packing. These three packings are known to crystallo-

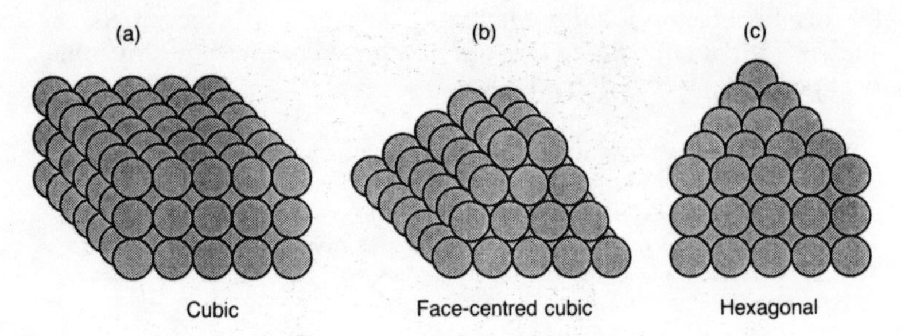

Figure 4.4 Kepler's three types of sphere-packing. (a) Cubic lattice. (b) Face-centred cubic lattice. (c) (Three-dimensional) hexagonal lattice

graphers as the cubic lattice, the face-centred cubic lattice, and the (three-dimensional) hexagonal lattice.

Robert Boyle, famous for his gas law, espoused a version of the Atomist theory:

> When many Corpuscles do so convene together as to compose any distinct body, as a Stone or a Mettal, then from their other Accidents (or Modes) ... there doth emerge a certain disposition or contrivance of parts in the whole, which we may call the Texture of it.

But Boyle disliked geometrical principles, apparently because he took the ideas of the ancients rather literally. He pointed out that the faces of dodecahedral gemstones such as garnet are rhombuses, not classical pentagons. To the modern mind, it's a slightly strange view to hold rhombi as non-geometrical! Elsewhere he noted that alum crystals can be octahedral, salt cubical, and saltpetre can form a prism. But he warned against placing too much faith in such simplicity: if two salts are mixed, they can produce crystals whose forms appear not to be any particular combination of those of the two ingredients. You begin to appreciate the difficulties that scientists were up against: there were so many observations to explain, and it wasn't clear which were fundamental.

Robert Hooke, remembered for his law on the stretching of springs, took up Kepler's explanation, showing that close-packed spheres can create innumerable crystalline shapes (Figure 4.5). He wrote:

> Had I the time and opportunity, I could make probable, that all these regular figures that are so conspicuously *various* and *curious*, and do so adorn and beautifie such multitudes of bodies ... arise onely from three or four several positions or posture of *Globular* particles.

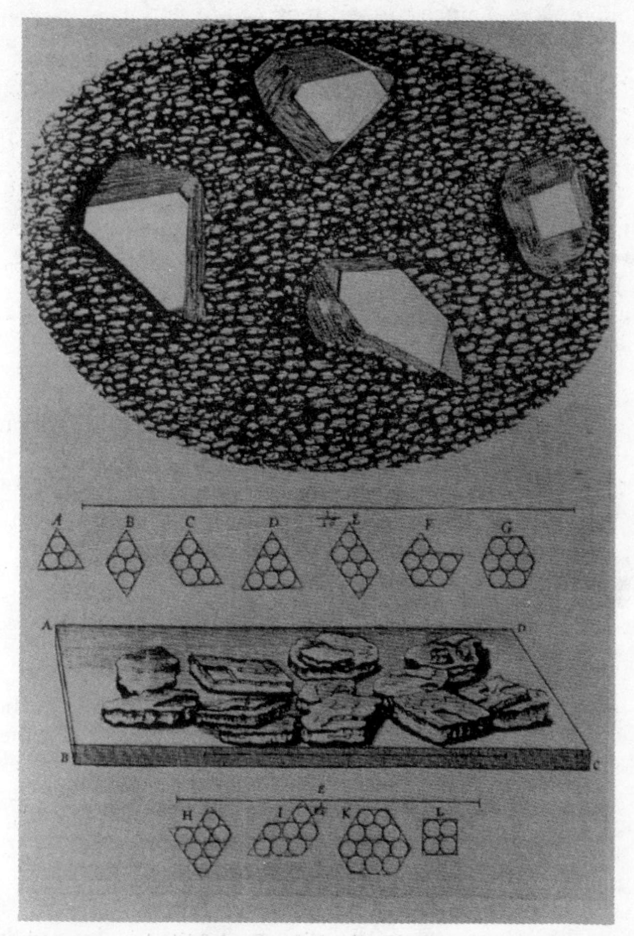

Figure 4.5 Hooke's theory of spherical particles, from his Micrographia *of 1665*

And he appealed to experimental evidence:

> Thus for instance may we find that the *Globular* bullets will of themselves, if put on an inclining plain so that they may run together, naturally run into a *triangular* order, composing all the variety of figures tha can be imagin'd to be made out of *aequilateral triangles*.

Scientists were groping their way towards a solution to the puzzle, painfully, but steadily. Their ideas until now had been largely qualitative. To get much further, they had to start making quantitative measurements. The spur was the problem of classification.

Oryctognosy

When scientists don't understand something, they often begin by listing all the possibilities. The approach is often dismissed as 'botany' or 'butterfly-collecting'; but botany (in that disparaging sense) does have its place. Linnaeus followed that approach when putting both botany and animal biology on sound foundations. Classification doesn't *explain* structure; but it does give a good idea of the patterns that underlie it. It's sobering to realize that the classification of crystals came *after* that of living creatures – partly because crystals are nowhere near as constant in form! In the absence of any good chemical theories, working with impure materials, it's hardly surprising that the pioneers had trouble. Here are two random extracts from a modern geological field guide.

> DIAMOND: octahedral, cube-shaped, dodecahedral crystals; often bulbous and striated faces.
> DIASPORE: Tabular, wide-columnar, foliaceous, acicular crystals; compact, lamellar, radial, tubular, scaly.

Imagine how far Linnaeus would have got with a classification such as

> GRYPHON: beaked, maned, quadrupedal, winged, feathered, furry, clawed.

And that's with the *modern* classification of minerals ...

No wonder that some scholars washed their hands of the whole business. 'The figure of salts does not appear to me to be of any utility,' said Georg Stahl, a prominent chemist. 'All the work of the crystallographers serves only to demonstrate that there is only variety everywhere where they suppose uniformity', Count Buffon echoed. The very word 'crystallographer' acquired the overtones that the scientifically literate nowadays associate with 'astrologer', 'ufologist', or 'pyramidologist'. But that had to change when Abraham Werner, professor of mineralogy at Freiburg, discovered how to slice through the web of complexity that had snared would-be classifiers. He called his system *oryctognosy*, the recognition of minerals from their external characteristics.

Werner was a pragmatist: he concentrated on what was practically feasible, and left subtler questions to others. Minerals, said Werner, have four types of property: those that we can observe directly from outside, their chemical composition, their physical properties (electrical and optical), and empirical ones (where they are found, how they relate to other minerals). Pay attention to the features that we can observe, said Werner; forget the rest. Such a simple idea, yet with it came the beginnings of order.

According to Werner, the main observable external features of minerals are colour, feel, coldness, weight, smell, taste and manner of aggregation. These in turn subdivide; for example manner of aggregation leads to shape, surface texture, external lustre, internal lustre, type of fracture, form of fragments, transparency, streak (scratch a board with it), hardness, solidity, adhesion to the tongue. Werner catalogued these properties for every mineral he could lay hands on. The catalogue could then be used in reverse, as a field manual, to identify a mineral by its observable properties.

Oryctognosy had one considerable advantage: it worked. It allowed mineralogists to conclude, with a considerable degree of certainty, that two apparently dissimilar crystals were different forms of the same mineral. And that paved the way for another brilliantly simple idea. Regularities of shape formed only a tiny part of Werner's system. But, because he could tell when two different crystals came from the same mineral, he could distinguish between what he called its primary form, and more complicated variations.

How? Easy! The primary forms were ... the simplest ones.

At first, Werner held that there are six primary forms: icosahedron, dodecahedron, prism, pyramid, table, and lens. Later he added the cube. He noted that the primary form can be modified by slicing bits off (a process that the geometers call *truncation*, see Figure 4.6):

> Galena has five kinds of crystals. ... The perfect cube is its first crystal form; when its corners are slightly truncated, we obtain the second crystal form. ... The fifth and last crystal form of galena ... is a perfect octahedron.

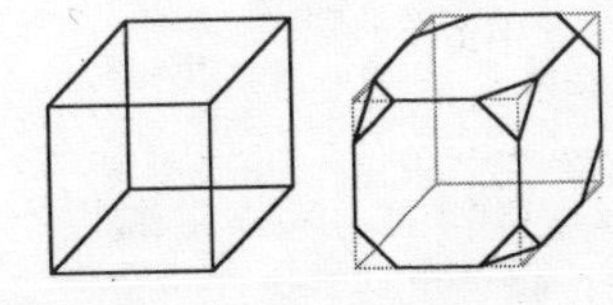
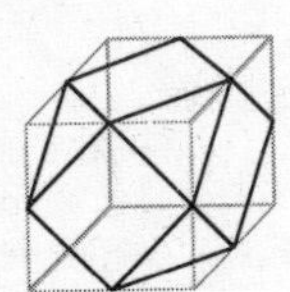
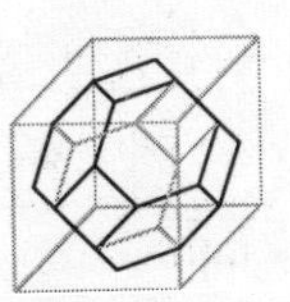
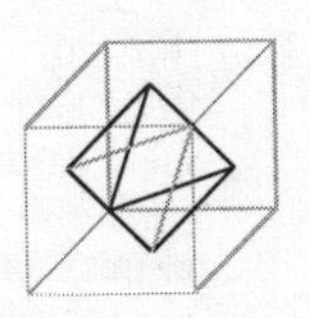

Figure 4.6 Different truncations of a cube: all are possible forms for a crystal of galena

Pragmatist to the last, he specifically did not believe that the primary forms are physical characteristics of the mineral. They were just a convenient, idealized version from which many actual forms could be deduced. He pointed out that the primary form of galena could as well be an octahedron as a cube; in the absence of any unique choice, why choose at all?

A New Angle

Independently, the notion of truncating simple forms was greatly elaborated by Jean Romé de l'Isle, whose four-volume treatise on crystallography appeared in 1783. For Romé de l'Isle, however, the simple forms were not mere ideals: they were real shapes to which minerals must, in some manner, conform. He described this hypothetical perfect form of each of 47 types of mineral. But more than that, he listed measurements of various angles: the angles at the corners of faces, the angles of the crystal axes (lines parallel to the edges of the crystal faces), and the angles between adjacent faces (or *interfacial angles* in the jargon). Not very accurately, mind you: for instance, he stated that certain angles in the faces of quartz were 'about 70° to 75°'. But it was a start.

Enter Christian Huygens. Huygens was one of the great founders of optics, and his interest in crystals was centred on a curious optical property of iceland spar. If you look through it from some directions, you see double images (even when sober): this is called 'double refraction'. In 1669 Erasmus Bartholinus had made accurate observations of the interfacial angles of this crystal. In his 1690 *Treatise on Light* Huygens begged to disagree slightly with Bartholinus's values – tacitly implying that those values must be *the same* in *all* crystals of iceland spar, since otherwise there would be no point to a disagreement. It was the first genuine mathematical regularity: the constancy of interfacial angles. The idea was already in the air: Steno and Romé de l'Isle were groping towards it, and it was *explicitly* recognized by Arnould Carageot, one of Romé de l'Isle's students:

> The author, a novice in crystallography ... was working at cleaving and modelling in clay life-size crystals ... he thought of cutting tentatively out of cardboard the angle that two of the faces formed with each other. When this angle had been cut, he was surprised to find the same angle on the two opposite faces, and so successively on the other faces of the same crystal. ... When the experiment was repeated on all rock crystals he had at hand, he recognized with satisfaction that the angles were constant.

We could debate who was the first discoverer of the principle; but it's the principle that counts. The interfacial angles of a given mineral are always the same.

Why is this constancy so important? It implies the existence of a *number* that can be *measured*, and which is characteristic of the mineral concerned, *independently* of the particular shape that the crystal takes. Knowing that number, the theoretical ideal form is immediate; and then – as Romé de l'Isle demonstrated at length – you can start from the ideal form, cut bits off according to his truncation rules, and reconstruct the form actually observed in the specimen. Constancy of the interfacial angle is the clue whereby the hidden ideal can be extracted from the far-from-ideal form found in nature. Recall Plato's image of the observed world as the distorted shadows of a pure ideal. It's easy to work out what shadow a given object will cast, knowing its shape and the positions of light source and screen; but constancy of the interfacial angle solved the 'inverse problem': deducing the underlying ideal from the imperfect shadows on the wall of Plato's cave.

Weaver's Son

At this point in its development, the fledgling science of crystallography found itself in possession of a penetrating theoretical idea, that matter is build from innumerable identical components – atoms – and a pivotal experimental discovery, that there are universal constants hidden in the apparent diversity of crystalline form. However, there was no strong mathematical connection between the two. Their synthesis was accomplished by the Abbé René Just Haüy, whose starting point was the observation that whenever a crystal of calcite is split into tiny pieces, then no matter what the original shape, all of the pieces have rhombic faces – lopsided squares. The reverse process must also be possible, and Torbern Bergman stated explicitly that all the observed crystal forms of calcite can be made by sticking these rhombohedral pieces together. Moreover, they aren't just added randomly: they are laid down flat on the existing faces as the crystal grows, like layers of tiles on a roof. As the layers grow, their sizes change in some regular manner; and the result is a flat-faced geometric solid.

Haüy, the son of a weaver, was born in 1743 in the village of Saint-Just-en-Chaussée. He was a bright student who chose a clerical career: he was ordained as a priest in 1770. Around that time he became interested in minerals, and started collecting them. Among them were calcite, whose crystals are hexagonal prisms, and iceland spar, which occurs as rhombohedra. Because of these two quite distinct crystal forms, they were considered to be different minerals, even though they were known to have similar chemical constituents. According to legend, Haüy accidentally dropped a beautiful hexagonal prism of calcite. As he began to clean up the mess he realized that the fragments were rhombohedra, exactly the same shape as complete crystals *or* fragments of iceland spar. Two very different geometric forms yielded the same shaped fragments when broken. Moreover, calcite and iceland spar had the same chemical constituents. It was not hard to conclude that the difference between the two crystal forms was merely the manner in which these identical components were assembled. 'Everything is revealed!' cried the priest in excitement. The story may well be apocryphal, but it's quite plausible. Even if it revealed everything to Haüy, what was new to him had been known to others for ten years or more. Equally, he did a lot more with it: over the next forty years he published more than a hundred technical papers elaborating 'his' theory of crystal structure.

He began with the ideas already presented: that tiny fragments of crystals have a form that depends only on the chemical nature of the mineral, not on the precise form in which it has crystallized. His method was to break crystals into very small fragments, with great care, and observe what shape occurred. Call this the *nucleus* of the crystal. He assumed that the atoms forming the mineral are of the same shape as the nucleus, but smaller. He checked that he could reconstruct the various crystal shapes of that mineral from their nuclei. For example, he showed that by stacking tiny cubes in a perfectly regular lattice on the faces of a larger cube, it is possible to build a rhombic dodecahedron (Figure 4.7). Alternative stacking rules lead to other forms. Building up complex designs from the repetition of identical elements is rather a natural idea for a weaver's son to have. Was Haüy subconsciously influenced by watching his father at work? We'll never know – but it's an intriguing thought.

By their very definition, nuclei have an important and unusual mathematical property. They are solids that can be stacked, in a perfectly regular manner, to fill space. This fact was observed, but its importance was not. For now we leave that particular thread dangling and pick it up in a later section.

Haüy's theory was simple, but it didn't remain that way for long, because he wasn't always able to produce just one shape for the nucleus of a given mineral. In compensation, however, he noticed that all mineral nuclei reduce to one of just six basic shapes:

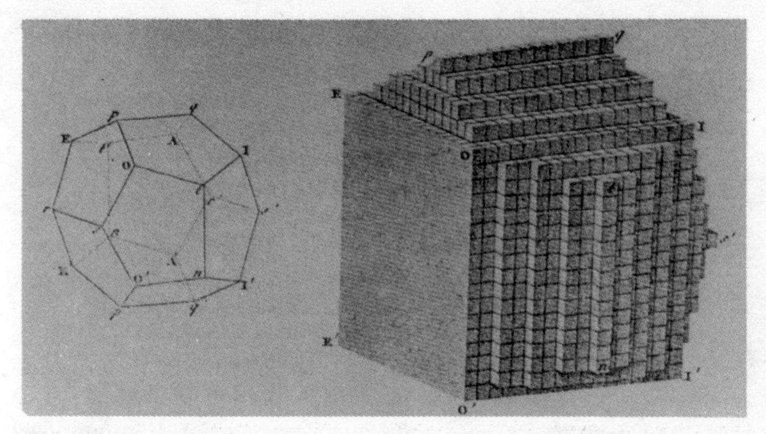

Figure 4.7 A rhombic dodecahedron built from cubes, from Häuy's Traité de Minéralogie *of 1801*

- parallelepiped (lopsided box);
- rhombic dodecahedron;
- triangular-faced dodecahedron;
- hexagonal prism;
- octahedron;
- tetrahedron.

By using the geometry of these basic forms, and one further assumption, Haüy was able to *calculate* the angles at the corners of the faces, and the interfacial angles. The assumption was that the plane along which calcite tended to split most easily met the crystal's main symmetry axis at *precisely* 45°. Despite all the strange angles measured by crystallographers, nobody had come up with anything as natural and simple as 45° before. This assumption led to a value of 101°32′20″ for the larger angle on the rhombic face of calcite, and 104°28′40″ for the interfacial angle. De la Hire's experimental result for the former, obtained in 1710, was 101°30′, differing by only two minutes of arc. This close agreement could scarcely be coincidence, and was felt to be strong evidence that Haüy's assumption was correct. This in turn meant that more accurate experiments ought to confirm his calculated figure.

Romé de l'Isle couldn't follow Haüy's geometry, and remained unconvinced: 'This example ought to put us on guard against these pretended geometrical demonstrations over which there is so much uproar.' Haüy, safe in the knowledge that his calculated angles agreed extremely closely with what was observed, ignored him. In 1802 William Wollaston made careful measurements, obtaining a figure of 45°23′ for the angle that Haüy thought must be exactly 45°. The priest objected on grounds of mathematical simplicity. For example, he pointed out that his figure of 45° led geometrically to a ratio of $\sqrt{3} : \sqrt{2}$ for the diagonals of the rhombic face of calcite, an elegant and simple value. The simplest value consistent with Wollaston's measurement was $\sqrt{111} : \sqrt{73}$. Anticipating a line of argument subsequently espoused by the great physicist Paul Dirac, Haüy asserted that he preferred a mathematically beautiful theory to one that agreed with observations.

We're keeping up our sleeve the fact that Haüy was wrong, Wollaston right – the modern figure is almost precisely Wollaston's. Haüy had overdone it: to a good idea, construction from identical basic components according to systematic principles, he had added a bad one – insistence upon too naive a kind of mathematical simplicity. This is always the danger with the Dirac position: what looks beautiful to us imperfect beings may seem ugly – or at least, irrelevant – to mother Nature.

Science, ever the opportunist, kept the good idea and discarded the bad one.

Crystal Symmetry

Despite the beautiful symmetry of many crystal specimens, especially those deemed 'perfect' (because symmetrical!), scientists had rather lost sight of the fact that crystal structure has strong features of symmetry. This missing ingredient was restored by Christian Weiss in 1804, in an article entitled 'A dynamic view of crystallization'. Weiss began with the idea that particles of matter need not just attract each other: they may also repel. All natural form, said Weiss, results from the interplay between attraction and repulsion of the basic constituents of matter. His reasoning was simple: if the only forces were attractive, all matter would condense to a single point. Something had to keep the basic particles apart once they got sufficiently close. In these terms, Weiss offered an elaborate and rather obscure explanation of the differences between liquids, gases, and crystals. But he also suggested that the forces might possess directional properties, leading to specific angles between coalescing particles; and that led him to the concept of the symmetry axes of a crystal.

Previous investigators had observed that many crystals possess an axis of symmetry. For example, a hexagonal prism has six-fold rotational symmetry about an axis perpendicular to its hexagonal faces (Figure 4.8). But everyone had assumed there was only *one* symmetry axis. Thus Haüy remarked that 'there are two sorts of symmetry, the perfect and the imperfect. In the perfect the right is symmetrical with the left and the top with the bottom, but in the imperfect, the top is not symmetrical with the bottom'.

Not so, said Weiss. Cubic crystals, for example, have not one, but *three* symmetry axes – the lines joining the mid-points of opposite faces. (A modern crystallographer would consider there to be many more, for example the long diagonals.) They are thus distinguished from other box-shaped crystals whose sides are of unequal length. The symmetry axes have physical importance as well: they determine important optical properties of the crystal. To Weiss, the crystallographic axes were the fundamental things; such surface features as interfacial angles were – well, superficial.

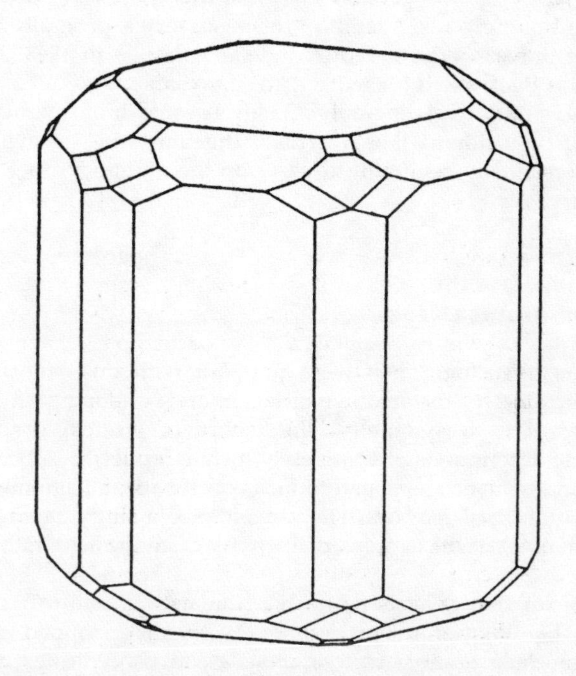

Figure 4.8 Crystals of beryl have hexagonal symmetry

Between 1811 and 1815 Weiss embarked upon a study of the mathematics of crystal symmetry. He classified crystals into two main classes: those with three axes at right angles to each other, and those with four axes, three being equally spaced at right angles to the fourth. He further subdivided these classes. For example, in the first class the lengths of the three axes can be equal; or two of them can be equal while the third is different; or all three can be different. In modern terms, Weiss had discovered the cubic, tetragonal, orthorhombic, and hexagonal systems of crystals. The analysis had flaws, but all kinds of things began to slot into place. At much the same time, Friedrich Mohs – who followed in Werner's footsteps at Freiburg – was also developing a classification of crystal structure based upon symmetry. Moreover, Mohs realized that there can exist families of crystal axes beyond those listed by Weiss – leading to what are now called the triclinic and monoclinic crystal systems. In modern language, Weiss and Mohs had discovered *point groups* – the groups of symmetries of crystal structure relative to a single point. In 1830 Johann Hessel gave a geometrical proof that there can exist only 32 different point groups. Credit for this discovery is generally given to Auguste Bravais, who obtained the same results in 1848. Hessel's paper was overlooked for more than fifty years.

In particular, crystallography finally became fully mathematical. Was it just coincidence that at virtually the same time, the mathematicians were finally beginning to develop the natural mathematics of symmetry?

The Mathematics Deepens

While the crystallographers were grappling with a nascent theory of crystal symmetry, the mathematicians were developing an entirely new approach to symmetry: the theory of groups. In place of geometric descriptions of symmetry, such as 'cube-shaped' or 'hexagonal', mathematicians began to focus on the idea of symmetry as a collection of *transformations*. The concept of symmetry became active rather than passive, expressed in terms of movement rather than form.

The belief that crystals *do* have mathematical features was confirmed when Werner, the inventor of oryctognosy, stopped worrying about the deep inner nature of crystals and concentrated on what could actually be detected and observed. The elaboration of an appropriate framework for those regularities came about in a similar manner: scientists stopped worrying about something that couldn't (at the time) be detected: the physical nature of the 'ultimate constituents' of crystals, and began to represent these constituents abstractly in drawings, as mere dots. Then they could think about the mathematics of regularly stacked dots without perpetually getting sidetracked by what the dots *were*.

Using abstract dots makes an immense difference to the questions you think are important; in particular it makes you clarify the underlying principles. For example, what do we mean by 'regularly stacked'? If you go back to Haüy's pictures you'll see that the nuclei are stacked in parallel rows. Such a pattern of dots is called a *lattice*. To create a lattice, pick three independent directions in space, and associate to each of them a fixed length. Draw one dot, and call it the origin. Then move away from the origin in steps whose sizes are those chosen distances and whose directions are parallel to the three chosen directions, placing a dot at each step. Repeat the entire process starting at each of the new dots, and continue until no further dots can be placed. The result is a kind of three-dimensional grid, which can be thought of as being packed full of copies of a single parallelepiped (three-dimensional analogue of a parallelogram) whose sides are parallel to the three chosen directions and have the chosen lengths. Alternatively, start by filling space with cubes as in Figure 4.9, put a dot at every corner, and then stretch the sides of the cubes by arbitrary amounts and tilt their edges to arbitary angles.

Lattices have some obvious symmetries, namely, translations along the three basic directions through the three associated distances. But they can sometimes have *more* symmetry. For example, a lattice composed of perfect cubes (Figure 4.9) also has all the symmetries of a cube: various rotations and reflections. The 'extra' symmetries of the lattice can be found by looking only at symmetries that fix the origin: these form the lattice's point group, a concept we have already mentioned.

For simplicity, let's think about the analogous situation in the plane. A lattice is then a repeating pattern of dots based upon parallelograms rather than parallelepipeds (Figure 4.10). There are four distinct possibilities for point groups. The generating parallelogram can be a square, a rectangle, a rhombus with angles of 60° and 120°, or a 'general' parallelogram. The point group correspondingly is:

- D_4: symmetries of a square;
- D_2: all symmetries of a rectangle;
- D_6: all symmetries of a regular hexagon;
- D_1: all symmetries of a parallelogram (namely rotation by 180°; and the identity).

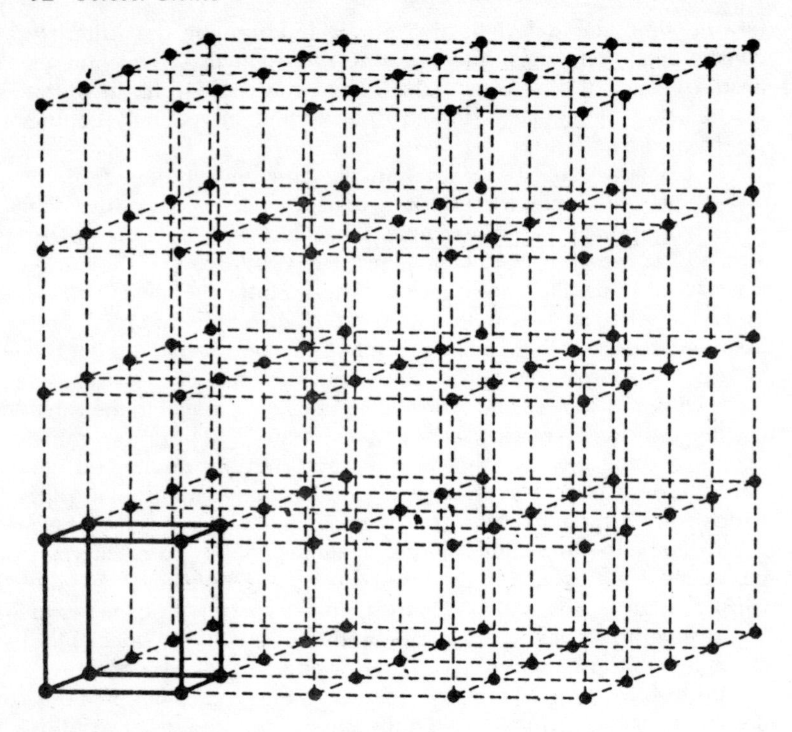

Figure 4.9 A cubic lattice and one of its fundamental cells

Different lattices can sometimes have the same point groups. In the plane this accident occurs only when the point group is D_2: see Figure 4.11. So not only is the form of the lattice, and that of its point group, important: so is the way that the two interact.

That is especially the case in three dimensions, where there are 7 types of point group and 14 distinct types of lattice, known as the *Bravais lattices*. They are shown in Figure 4.12, and form the basis of the modern classification of crystals according to their symmetries.

It's all so much simpler when you concentrate on the basics.

These two classifications concern the point symmetry of the lattice, and the coarse features of its overall structure. In three dimensions these ingredients interact in much more intricate ways than they do in the plane. By combining the point group with the lattice, it turns out that there are precisely 230 different *space groups* – symmetry

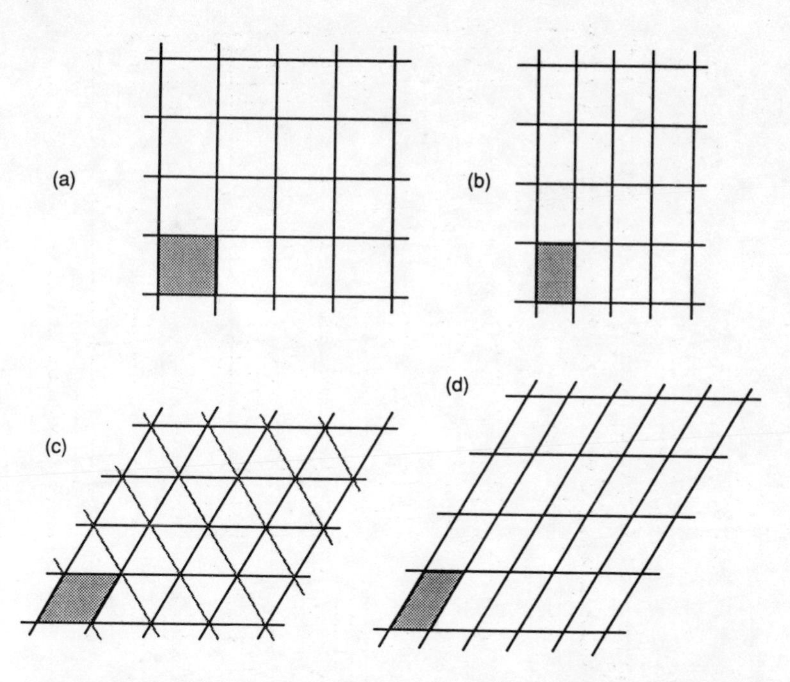

Figure 4.10 The four kinds of two-dimensional lattice. (a) Square. (b) Rectangular. (c) Hexagonal. (d) General

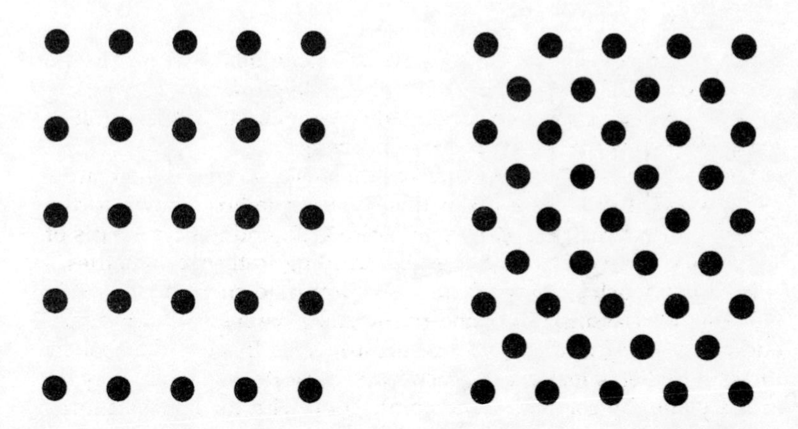

Figure 4.11 Different Lattices with the same point group D_2

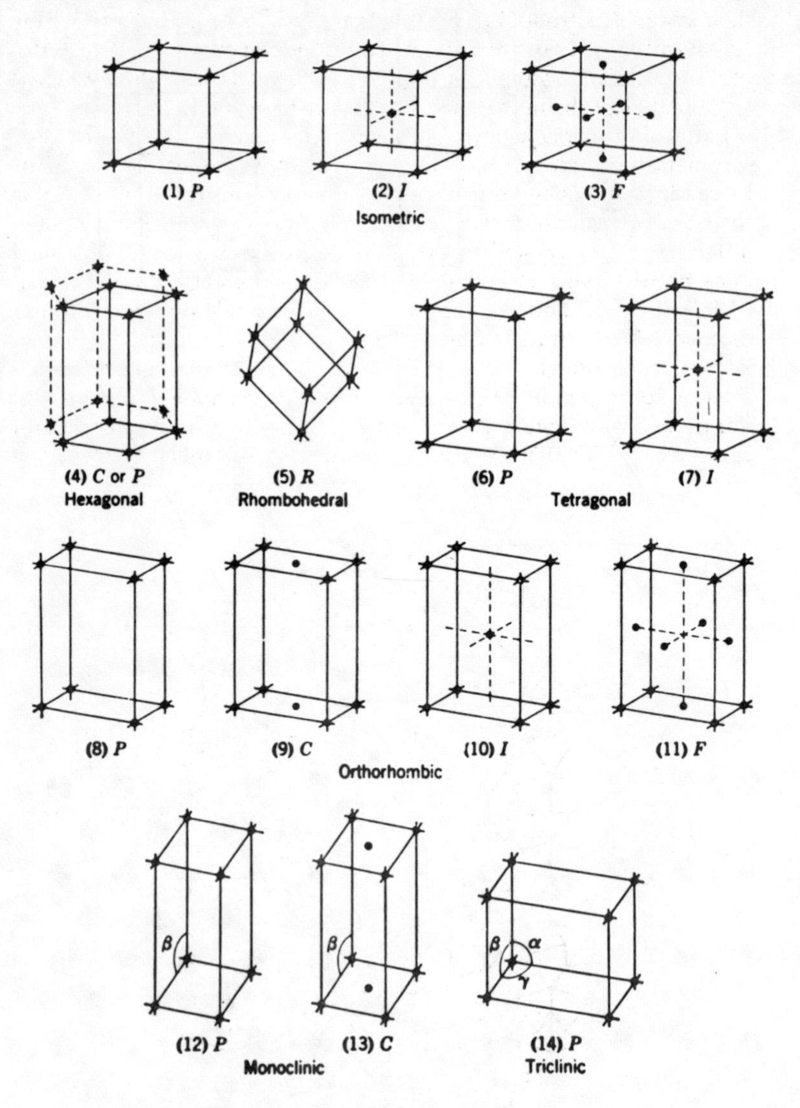

Figure 4.12 The 14 Bravais lattices in three dimensions

types of lattices, distinguished not just by the abstract structure of their symmetry group, but by how that group acts on the lattice. This classification was obtained somewhat independently by Evgraf Fedorov and Arthur Schoenflies in 1891. William Barlow also obtained large parts of it using geometric trial-and-error methods.

Lattices in the plane or three-dimensional space can have rotational symmetries of orders 2, 3, 4, and 6, but no other values. In particular, there cannot be five-fold rotational symmetry. This result is known as the *crystallographic restriction*. Very recently, attention has turned to a different type of 'crystalline' structure, known as a *quasicrystal*, which does not involve a basic lattice at all. Quasicrystals 'almost repeat' their structure, and *can* have axes of five-fold symmetry. Their discovery, as abstract mathematical objects, is due to the mathematical physicist Roger Penrose, who – purely for recreation – invented a set of tiles based upon pentagonal shapes which tile the plane, but not periodically (Figure 4.13). Oddly enough, in 1619 Kepler's mind was moving in similar directions (Figure 4.14), but this particular idea

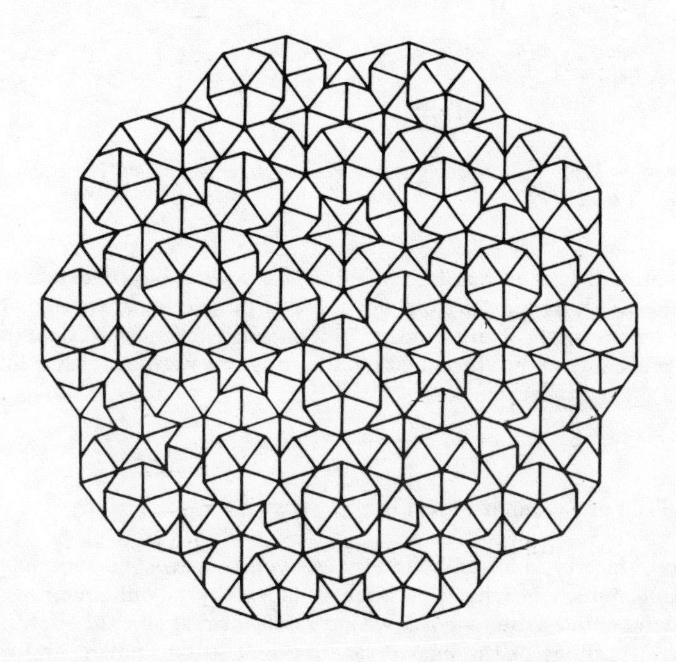

Figure 4.13 A Penrose pattern: two types of tile, arranged aperiodically, with approximate five-fold symmetry

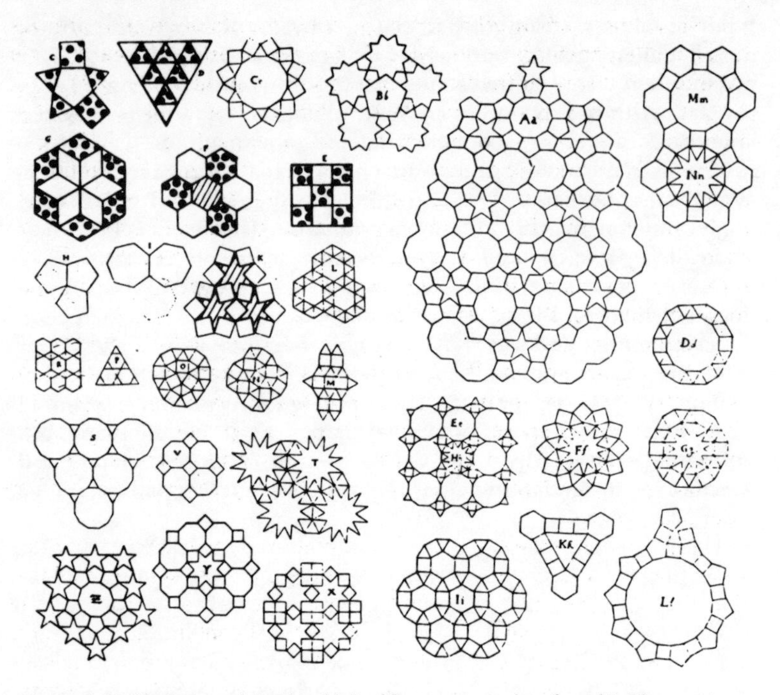

Figure 4.14 Kepler's anticipation (Aa) of Penrose patterns, from his Harmonices Mundi *of 1619*

got shunted into an intellectual siding for several hundred years. It would divert us too far from our topic to pursue these strange and wonderful objects at any length; but their existence makes it clear that the prevailing group-theoretic paradigm is not necessarily the final word on crystal structure.

The Role of Symmetry-breaking

A lot of history, a lot of symmetry. As yet, however, no symmetry-*breaking*, which is what this book is (allegedly) about. Symmetry-breaking enters when we ask why crystals form at all. Although the general outlines of the answer are thought to be known, and are supported by computer calculations, a rigorous proof that the tale we are about to tell is really true remains as a challenge to future mathematicians and physicists.

First, we now know what the basic constituents of crystals are. As for all matter, they are atoms. In 1915 Lawrence Bragg developed the technique of x-ray diffraction, which bounces beams of x-rays off a crystal and watches how the waves interfere, and showed that the atoms of a perfect crystal are arranged in a lattice (Figure 4.15). The mathematical structure of the lattice, and in particular its symmetries, affects the physical properties of the crystal – because, not surprisingly, the way in which the atoms are arranged makes a difference. From this point of view a crystal is just a huge molecule built by repeating the same basic unit over and over again. Finally the mathematics and the physics come together.

Imagine a hot gas of identical atoms, condensing to form matter. As a mathematical system, such a gas has an enormous amount of symmetry. The equations that describe it are unchanged under all rigid motions of space; and also under all permutations of the individual atoms. If you take a gas, reach in, and swap a couple of identical atoms around, you can't tell the difference. That's what 'identical' means.

The atoms attract each other by interatomic forces, so as the gas cools they tend to clump together. What is the most symmetric form that they can take up? It's a mathematical fiction, a solution of the equations, but one ruled out by other physical constraints: *put all the atoms in the same place*. This state of the 'gas' has all of the symmetries just mentioned: it's the unique solution of the equations that doesn't break symmetry.

However, real atoms can't occupy the same space as each other – firstly, because the inter-atomic forces become repulsive when the atoms get close together, and secondly because it would be ridiculous if two atoms really could occupy the same space. This may or may not be the reason why the deity provided repulsive forces at small distances, who knows? If we build in those repulsive short-range forces, anticipated by Christian Weiss on identical grounds to those just rehearsed, then this impossibility effectively shows up in the mathematics as the *instability* of the fully symmetric solution.

We've encountered this phenomenon before. What happens when a fully symmetric state of a highly symmetric system becomes unstable? Of course, the symmetry *breaks*, and the system takes up a stable state with less symmetry; but usually, as we have seen, quite a lot of symmetry – indeed the more the merrier. Crystal lattices have a great deal of symmetry, but they are not invariant under all rigid motions – only under those that are symmetries of the lattice. Moreover, the atoms in a lattice do not coincide: the dots are separated, and spread out evenly. So on general grounds of symmetry-breaking, a lattice is a *highly plausible* structure for a solidifying gas.

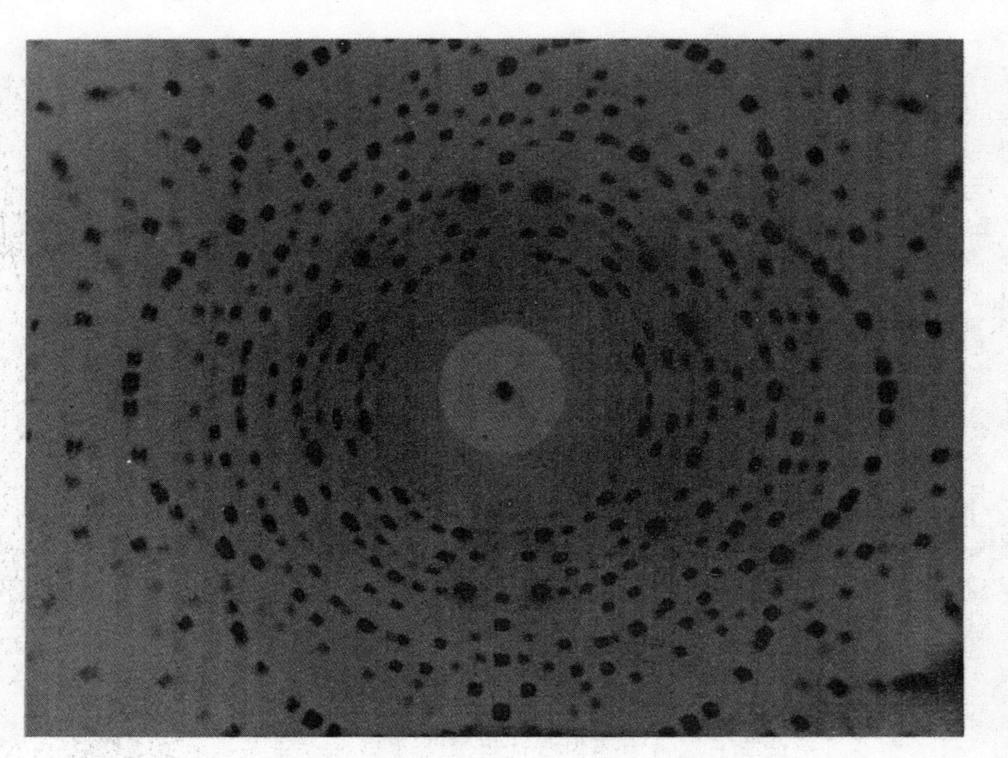

Figure 4.15 X-ray diffraction pattern from a crystal of beryl. Note the hexagonal symmetry

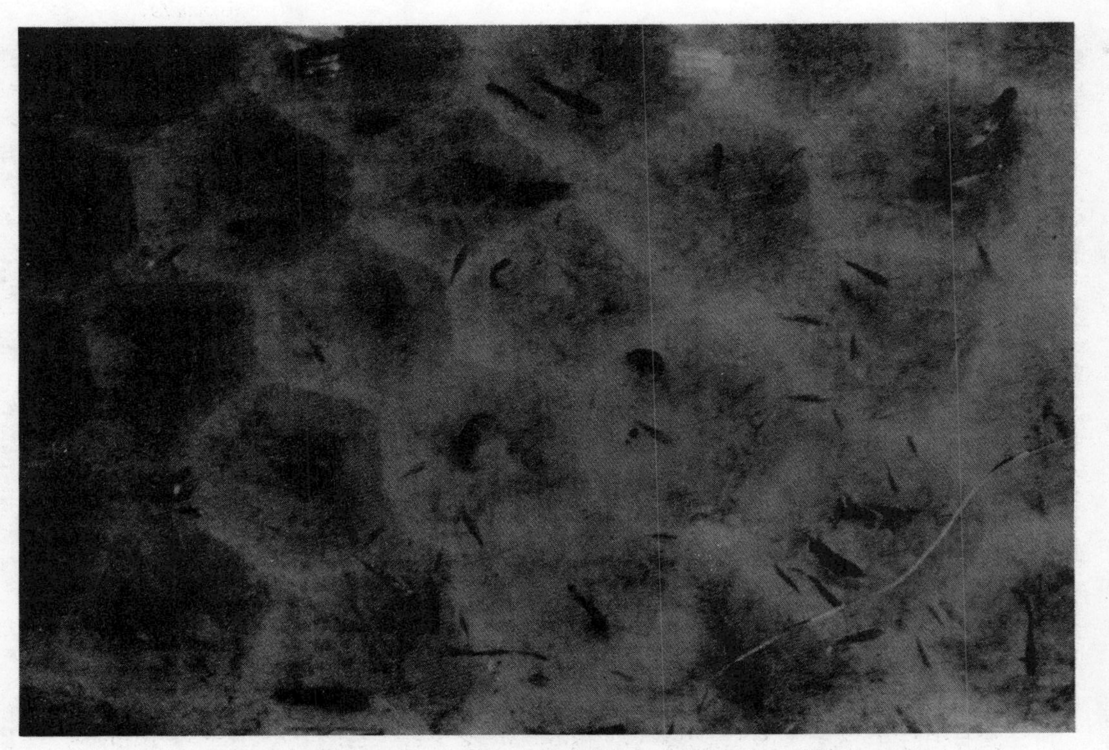

Figure 4.16 Hexagonal lattice formed from fish territories in Lake Huron

What's lacking is any rigorous proof that the crystal form is itself stable. Computer calculations suggest that it should be, but nobody has yet proved it. Indeed, only recently has it been proved that *gases* exist, in the sense that the physical equations for a system of many atoms has solutions in which the atoms are widely separated and largely independent of each other. The liquid state, where the atoms interact more strongly, but still with considerable freedom, has eluded all attacks. The crystalline state, where the interactions between atoms are very strong, appears totally intractable.

It's not just atoms that behave in this way. In Lake Huron there is a species of tiny fish that displays strong territorial instincts. The fish repel all invaders who approach within a distance of around 15 cm. On the other hand, population pressure tends to force the fish close together. This is closely analogous to what happens in a crystal: the fish play the role of atoms, population pressure is the attractive force, and territorial defence is the repulsive short-range force. So, by our general arguments, the fish ought to take up a crystalline state.

What nonsense! How can *fish* form a crystal?

Ah, but they do. The territories of the fish tend to pack together in a close-to-perfect hexagonal lattice (Figure 4.16). It's even clear why. Let's take a more abstract viewpoint. We'll approach the question by considering a population of identical circles in the plane, attracted to each other by forces beyond their control, but unable to interpenetrate (a condition equivalent to the existence of short-range repulsive forces). Start with the simplest case of two circles. Because of attraction, they bump together, but they can't interpenetrate, so they stop at that point. We now have a stable system of two touching circles. Moreover, that system has some symmetries: you can rotate it through 180°, and also reflect it about the line joining the centres of the two circles (Figure 4.17a).

Now add a third circle. It is attracted towards the two already in contact, and eventually bumps into one of them. Its stablest position is when it is in contact with them both. See Figure 4.17b. In any other position (with one exception) it will roll around until it reaches this stable position. The exception is when all three circles are in an exact straight line, but this is an unstable state so we may safely ignore it. The system of three circles now has reached a unique stable state, and it has quite a lot of symmetry: namely, all the symmetries of an equilateral triangle. Moreover, it has formed the nucleus, in Haüy's sense, of a hexagonal lattice, and as other circles settle on to it, they will tend to roll into position on lattice points. You can see exactly this sort of behaviour in bubbles on the surface of a mug of beer; and Robert Boyle gave pretty much this explanation in his remarks,

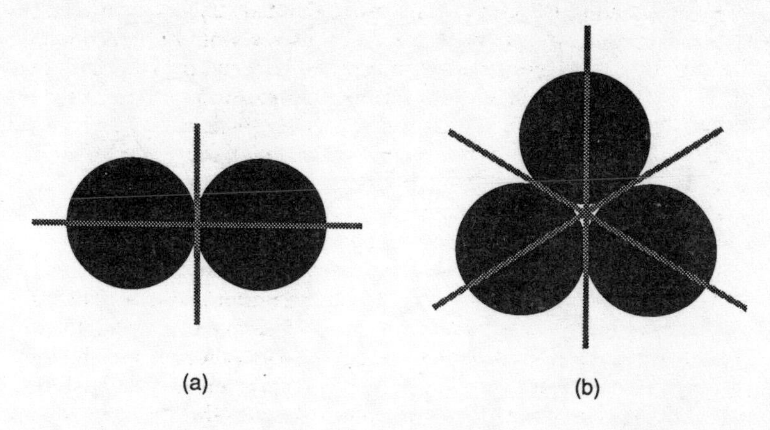

Figure 4.17 Automatic symmetry in the packing of circles. (a) Two circles. (b) Three circles

quoted earlier, about the way globular bullets pack together in a hexagonal lattice.

Indeed the mathematician Fejes Toth proved that the hexagonal lattice is the most efficient method for packing together identical circles, in the sense that it gets the largest number of them into the smallest space. The same goes for fish – and for atoms. This is analogous to minimizing their total energy relative to attractive and repulsive forces, so again we have evidence for the stability of a crystal lattice.

Why aren't *all* crystal lattices hexagonal, then? The answer is that other lattices can occur if the population doesn't consist of identical units, which is commonly the case, since most crystals involve several types of atom (Figure 4.18). Moreover, even identical atoms can exert different forces in different directions, which may cause them to pack in other ways.

What about three dimensions? Now we have to pack spheres rather than circles. In 1611 Kepler conjectured that the most efficient sphere-packing is a lattice packing, namely the face-centred cubic lattice. Although not as notorious as Fermat's Last Theorem, this *Kepler Problem* predates it, and has caused at least as many headaches for the mathematical profession. Carl Friedrich Gauss proved that among *lattice* packings, the face-centred cubic arrangement is best. But why should the best packing be a lattice at all? It may seem unlikely that an irregular jumble of spheres could fill space better

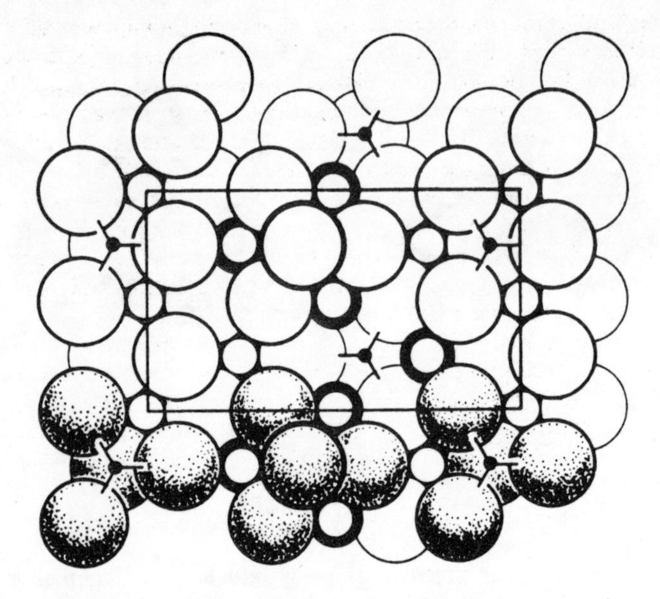

Figure 4.18 A crystal of olivine, involving several types of atom

than a regular arrangement, but it's not obvious how to prove such a statement. Moreover, it's actually *false* for packings in 1,000-dimensional space, where random arrangements can fill space better than any lattice! As we go to press, the American mathematician Wu-Yi Hsiang has just announced a solution of the Kepler problem. If his proof turns out to be correct, then we'll know for certain that the face-centred cubic lattice is not only better than all other lattices: it's better than any other arrangement whatsoever, regular or random. That would be strong evidence in favour of the theory that crystal structure is due to symmetry-breaking in a system composed of a large number of identical atoms.

One final piece of the jigsaw puzzle. Let us suppose, for the sake of argument, that systems of a moderate number of atoms tend to pack together in a unique way. The Russian mathematician Boris Delone has proved that if an arrangement of atoms looks 'the same' in a suitably large sphere surrounding each atom, then it must have perfect lattice symmetry. A sphere is 'suitable' provided its radius is six times as big as the distance to the nearest atom: it doesn't have to be *huge*. Now our assumption of uniqueness means that the configuration surrounding each atom *must* be idential in all cases, so Delone's

Theorem implies that a lattice structure must occur. Again, this isn't a rigorous argument: we haven't specified how large the unique clumps must be, or indeed explained why uniqueness is likely; moreover, by concentrating on moderate-sized clumps, we're neglecting interactions between widely separated atoms altogether. Nevertheless, we begin to see *why* the global regularity of a lattice structure might be caused by purely local interactions between atoms.

Next time you look at a diamond, don't think about its monetary value. Think about its immense value to science as the source, along with its fellow crystals, of some of the most fascinating branches of physics, chemistry, and mathematics that the world has ever seen. Think of all the marvels, from medical CAT scanners to digital watches, that have come from it. Above all, ponder the fact that without symmetry-breaking, diamonds wouldn't be forever. They wouldn't *be* at all.

5

Striped Water

> A hollow, and a solid cylinder might be so mounted as to admit of being turned with different uniform angular velocities round their common axis, which is supposed to be vertical. If both cylinders are turned, they ought to be turned in opposite directions, if only one, it ought to be the outer one; for if the inner one were made to revolve too fast, the fluid near it would have a tendency to fly outwards in consequence of the centrifugal force, and eddies would be produced.
> Sir George Gabriel Stokes, *Transactions of the Cambridge Philosophical Society* 1848

At the heart of our story lie the relations between mathematics and nature, modelling and reality, theory and experiment. The dynamics of fluid flow provides an ideal testing-ground for questions about such relationships. Here we take a look at two famous types of fluid flow and examine the patterns that they can produce. Both problems have stimulated a great deal of theoretical work on symmetry-breaking. Both are laboratory systems, although one of them, Bénard convection, also has applications to weather systems and to geology. The other, the Couette–Taylor experiment, was originally designed to provide a simple, patternless flow, but has turned out to generate a baffling range of patterns. At least forty different types of flow have been distinguished in this system. Only recently has it been realized that virtually all of them are related to the symmetry of the apparatus.

Shear Delight

In 1888 the French scientist M. Maurice Couette wanted to study shear flows, in which layers of fluid slide past each other. Shear flows are important in a variety of applications: they occur in the flow of fluid through pipes, in the atmosphere, around sails and, perhaps most importantly, around airfoils. The simplest mathematical model of shear flow is fluid between two moving planes. The fluid near the planes sticks to them, so the layers of fluid are forced to slide past each other. This model isn't much good for experiments, because: laboratory technicians don't have a ready supply of infinite planes, nor of the infinite quantity of fluid needed to fill the gap. Even using finite planes, it's difficult to keep the sideways motion going for long enough to obtain useful measurements: the planes would make a smart exit from the laboratory and disappear rapidly from view.

What's really needed is something that not only remains within the laboratory, but stays in the same spot on the bench. Couette used two cylinders, one inside the other. He filled the space between with fluid, and slowly rotated the inner cylinder to create the shear (Figure 5.1). Because rotation, unlike translation, can be continued indefinitely, the shear flow can now be studied at leisure. Couette was not the first to invent this apparatus. Isaac Newton discussed the motion of fluid between revolving cylinders in Book II of his *Principia* in 1687;

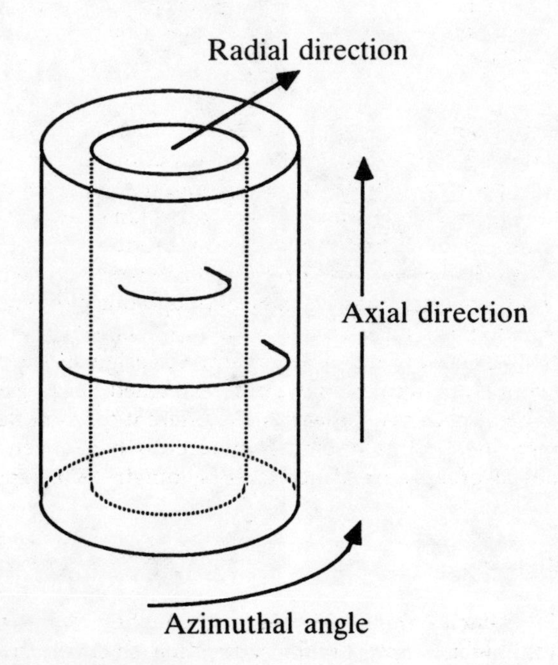

Figure 5.1 Couette's cylinders

a description quoted from Sir George Stokes in 1848 decorates the beginning of this chapter; Max Margules, an Austrian meterologist, proposed using it to measure viscosity in 1881; and Arnulph Mallock, assistant to Lord Rayleigh, actually used the apparatus to measure the viscosity of water in 1888 – the year in which Couette began his investigations.

Shear flow between concentric cyclinders has all sorts of features that are of interest to a fluid dynamicist, but from the point of view of pattern it lacks appeal, because it seems not to have any (Figure 5.2a). It's a featureless, smooth flow that looks pretty much the same at all points of the cylinder. This simplicity was important to Couette, making it easy for him to perform quantitative studies; but if there were nothing more to Couette flow, this chapter would have been about something else. Fortunately for us, Couette's system generates a whole range of remarkable flow-patterns, in addition to the relatively boring shear flow for which he designed it.

Figure 5.2 Patterned flows in the Couette–Taylor system. (a) Couette flow. (b) Taylor vortices. (c) Wavy vortices. (d) Turbulent Taylor vortices. (e) Spirals

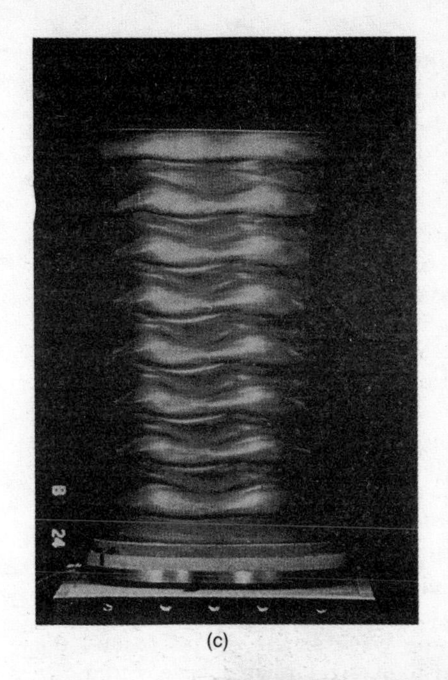

(c)

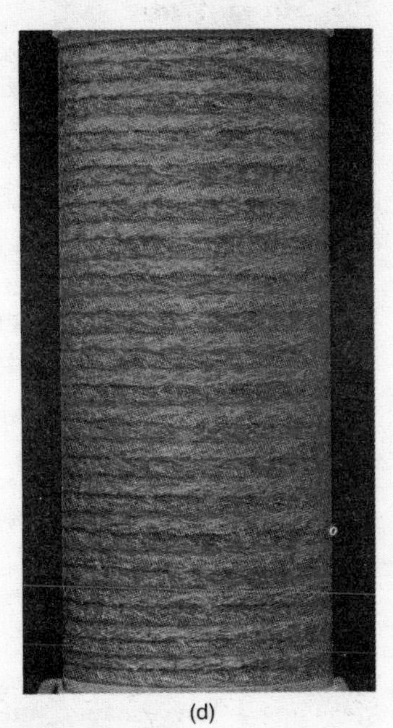

(d)

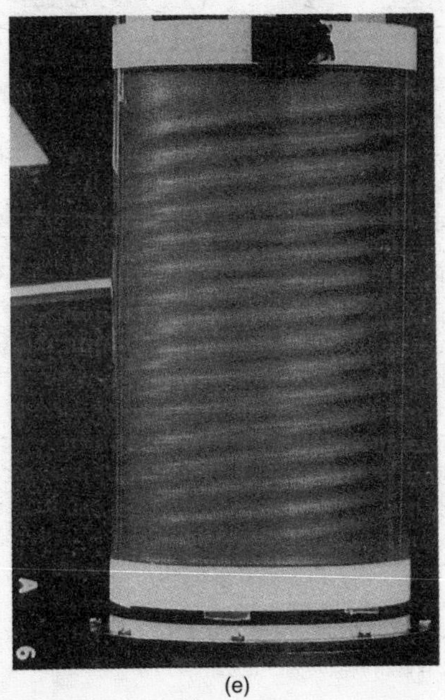

(e)

The liberation of the flow-patterns imprisoned between Couette's cylinders began with the British applied mathematician Geoffrey Ingram Taylor in 1923. Taylor noticed that if you speed up the inner cylinder, then the flow stops being uniform and boring. Instead, it breaks up into regular and repetitive layers of vortices (Figure 5.2b), like a stack of doughnuts or polo mints (US: lifesavers). In each vortex the fluid flows in paths that spiral round a torus, and the vortices come in pairs, with the direction of the spiral reversing from one cylinder to the next. The boundaries between these vortices are flat, and only by observing closely can it be seen that the fluid in the vortices isn't stationary. It's a 'steady flow' – which doesn't mean that the fluid isn't moving, but that the *velocity* at any point remains the same as time passes.

Ponies and Pandora

Speed up the inner cylinder still further: what happens? The flow ceases to be steady, and the fluid velocity at a given place starts to vary with time. You still see vortices, but the boundary between them develops a wavy shape, and the wave slowly rotates around the cylinder. This state – not surprisingly – is known as *wavy vortex flow* (Figure 5.2c)

Anything that works twice should be tried again, so let's speed up the cylinder still more. Now the wavy boundary oscillates up and down as well as rotating, rather like ponies on a merry-go-round. The traditional name is more severe: *modulated wavy vortices*. (However, a related dynamical effect is sometimes christened a POM, which stands for 'Ponies On a Merry-go-round'.) Keep speeding the cylinder up ... you see a more and more rapid sequence of less and less distinguishable changes, until the flow becomes turbulent, though still wavy. Call this *wavy turbulence*. If the speed is increased even further, there's a surprise: the layered structure of Taylor vortices re-forms, but superimposed, as it were, on the turbulence. This state is known as *turbulent Taylor vortices* (Figure 5.2a), and we'll have more to say about it in chapter 9. Still higher speeds lead to featureless turbulence, looking much the same everywhere.

It's a complex and puzzling series of changes – but there's quite a nice pattern to it, the so-called 'main sequence' of bifurcations:

- Couette flow;
- Taylor vortex flow;
- wavy vortex flow;

- modulated wavy vortices;
- wavy turbulence;
- turbulent Taylor vortices;
- featureless turbulence.

A rich variety of flows ... but the Couette–Taylor system has scarcely begun to perform its repertoire of tricks. Experimentalists such as Don Coles at Caltech in the 1960s and the group under Harry Swinney at Austin, Texas in the 1980s modified the apparatus so that the outer cylinder too can rotate. You might think this is pointless: surely only the *difference* in the speeds of inner and outer cylinder matters? Not so! The reason is that the cylinders are curved rather than flat. A rotating body is subject to centrifugal force, flinging it outwards radially. It's easy to see that this force doesn't depend just on the difference between the speeds, by considering extreme cases. If both speeds are zero then the cylinders remain fixed, and there's no centrifugal force. If the entire apparatus is placed on a turntable and rotated, then both cylinders turn at the same rate and the difference in speeds remains zero; but since the entire apparatus is rotating, centrifugal forces now arise.

In fact the Couette–Taylor apparatus is remarkably sensitive to the individual speeds of the two cylinders. This was apparent only in subtle ways in the first experiments that rotated both cylinders, which rotated them in the same direction. Differences in the flows here were mainly quantitative, although new states labelled twisted vortices and braided vortices were observed. But the sensitivity appears in more dramatic fashion when the outer cylinder is rotated in the opposite direction to the inner. Now, instead of the familiar Taylor vortices, Couette flow changes into a pattern of helical *spirals*, like those on a barber's pole (Figure 5.2e). They behave just like the spirals on a barber's pole, too, appearing to move along it when in fact they're rotating. It's quite disturbing to watch spiral flow – you sit there waiting for the fluid to come out of the top of the apparatus, and feel somewhat foolish.

When *both* speeds are varied, Pandora's box falls open, and almost as many flow patterns escape as did noxious insects in the Greek myth. There are wavy spirals and interpenetrating spirals, modulated wavy spirals, wavy inflow and wavy outflow boundaries, even spiral turbulence. A diagram published in 1986 by David Andereck, S. S. Liu, and Harry Swinney, exhibits a veritable zoo of patterns, which depend on the speeds in a very complicated and puzzling manner (Figure 5.3). Don't imagine that it's easy to obtain such a diagram: each experimental data-point can take several hours of careful experi-

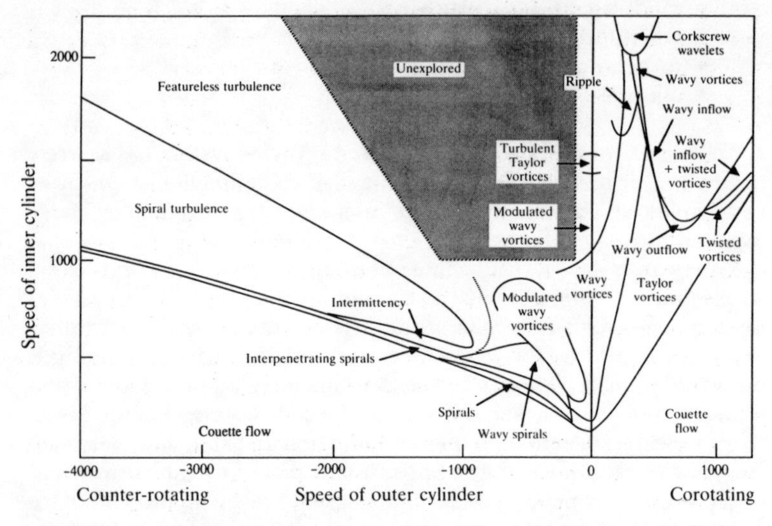

Figure 5.3 Experimental diagram showing how different combinations of cylinder speeds yield different flows.

ment, because fluid flows take a long time to settle down. In fact things are even more complicated than the diagram suggests, because the positions of the boundaries between different flow-patterns actually depend on the manner in which the speeds are varied. For example it can make a difference if the inner cylinder is brought up to speed before the outer, or after it.

What on Earth is going on?

Symmetry Rules, OK?

Over the last five years mathematicians have begun to understand the range of patterns observed in Couette–Taylor flow by first seeking an overall qualitative skeleton for the process of pattern-formation, and then fleshing it out with quantitative detail. In other words, the idea is first to pin down in general terms what kinds of pattern can happen, and then calculate which ones actually do. This idea was first exploited systematically by Pascal Chossat and Gérard Iooss at Nice around 1984, and virtually all of the states in Couette–Taylor flow, perhaps even the turbulent ones, can now be understood in terms of the symmetries of the apparatus. What *are* the symmetries of the Couette–Taylor apparatus? There are several; and one of the most important isn't really there! First, those that *are*. The most obvious are rotations. If the entire apparatus is rotated about its axis, everything looks exactly the same. So rotations about the axis are symmetries. There's also a top-bottom reflection. On the other hand, left-right reflection isn't a symmetry of the Couette–Taylor system, although it is a symmetry of the static cylinder. The difference is that left-right reflection reverses the direction of rotation, so it's not a dynamic symmetry. (The easiest way to see this is to imagine the apparatus placed next to a vertical mirror: if it spins clockwise, then its mirror image spins anticlockwise. In contrast, imagine it placed on top of a horizontal mirror: now cylinder and image spin in the same direction.)

The symmetry that isn't really there is translation in the vertical direction. Undaunted by the facts, Taylor devised a theory in which this symmetry *is* there, modelling the apparatus by an infinitely long cylinder. This makes the mathematics simpler, because an infinite cylinder doesn't have any ends, and ends cause all sorts of mathematical problems – no end of them, indeed – even though physically they're essential to stop the fluid leaking out.

More precisely, there's a surrogate for the ends, known as *periodic boundary conditions*. Imagine the infinite cylinder to be constructed out of infinitely many identical finite cylinders, stacked one on top of the other; and study only dynamics that involves the same flow-pattern in each finite component. Effectively this 'wraps round' the top of the actual finite cylinder and glues it on to the bottom (but without introducing any bending!). The rationale for this approximation (apart from the fact that it seems to work) is that all the patterns we want to study, like Taylor vortices, are periodic in the vertical direction; moreover, a single vortex cell occupies a fairly short segment in a long cylinder. Neglecting end effects by removing the ends to infinity shouldn't make too much difference to the flow near the middle. So the infinite cylinder model is fine for spatially periodic flows near the middle of relatively long cylinders. It definitely breaks down if the cylinder is short. Moreover, ends do exert various more-or-less subtle effects even in long cylinders, so you have to be careful.

There's an additional symmetry of a rather different nature, so simple that you may not notice it. What is it? Soon all will be revealed . . .

Patterns of Broken Symmetry

The symmetries just listed (including the translational 'almost' symmetries and the mystery symmetry yet to be revealed) are exactly the ones that we need to capture the observed patterns in the Couette–Taylor experiment.

The 'boring' Couette flow may be featureless, but that's not due to its lack of symmetry. On the contrary, it's due to its excessive degree of symmetry: it's unchanged by every single one of the symmetries listed. Couette flow has the same symmetry as the entire apparatus: there's no symmetry-breaking (and no pattern!). Rotate it, reflect it, translate it: it still looks uniform and featureless. Human psychology strikes again: *too much* symmetry isn't perceived as pattern.

Taylor vortices, on the other hand, break the translational symmetry. If you take a vortex pattern and move it a little bit up the cylinder, you can tell it has been moved because the boundaries between vortices are no longer in the same place. But if you move the pattern up by precisely the height of a vortex-pair (recall that the circulation direction of the flow reverses from one vortex to the next, so you have to step two vortices up or down to make the flow directions match), the flow looks exactly the same as before. So the translational symmetry breaks from the group of all translations to a subgroup, translation through integer multiples of the height of a vortex-pair. The rotational and reflectional symmetries of the apparatus are unbroken. Rotations don't affect the flow-pattern because the vortices look the same all the way around the cylinder. And provided we choose the 'correct' reflection – for example reflection in the boundary plane between neighbouring vortices – then the pattern also looks the same before and after.

Wavy vortices are far more subtle. They retain the discrete vertical translational symmetries of Taylor vortices, but they break both the reflectional and the rotational symmetries. However, these symmetries aren't totally lost. A reflection changes the shape of the wavy boundary between adjacent vortices; but it can be restored to its original position by a further rotation. This is a rotational version of a glide reflection. From the totality of all combinations of rotations and reflections, only the glide reflections – mixtures of the two – survive.

Typically, the boundary between wavy vortices consists of four or five identical waves. So, if you rotate the apparatus through one-quarter or one-fifth of a turn, the pattern repeats precisely. Again the group of all rotations is replaced by the subgroup formed by all multiples of some fixed rotation (90° or 72° respectively). There's yet another symmetry of wavy vortices, the 'mystery' symmetry mentioned above. Unlike Taylor vortices, wavy vortices aren't a steady flow, but a time-periodic one. The mystery symmetry is time translation. As we explained earlier, a steady state is unchanged by arbitrary translations of time; a time-periodic state is unchanged only by time translations through integer multiples of its period. If you stand opposite one point of the Couette–Taylor apparatus and observe wavy vortex flow 'rotating' past your eyes, then after the appropriate period of time, what you see will be identical with what you saw originally.

But even this isn't the final symmetry! The wavy vortex state is a 'rotating wave': the waveform appears to rotate as if rigid. If you revolve around the cylinder, in a so-called 'moving frame', at the same speed as the wave, what you see will be a steady state. The symmetry involved here is a mixture of space and time, because it takes time for the moving wave to rotate to a new position. You must rotate the apparatus through some angle, and also wait the correct amount of time, to observe the identical waveform.

Wavy vortices actually have quite a lot of symmetry. It just takes a trained eye to notice.

Symmetry considerations shed useful light on the structure of the 'main sequence'. Couette flow has full symmetry. Taylor vortex flow breaks some of it (most of the vertical translations). Wavy vortices break more. Modulated wavy vortices break most of the mixed spatio-temporal symmetries of wavy vortices: they become time-dependent in a much more complicated manner. Finally, the turbulent states have no obvious pattern, so presumably no symmetry at all.

Presumably ... but what about turbulent Taylor vortices? They may not have any symmetry, but they do have a sort-of pattern. We'll see in chapter 9 that even the turbulent states may have their own kind of symmetry. But for now, the essential point is this: the main sequence is a series of steps, at each of which symmetry is lost, until by the end it has all disappeared. It's a symmetry-breaking cascade.

That deals with the main sequence; but thanks to Swinney's group, there are other patterns to consider. The most fundamental, spiral flow, has vertical translational symmetries, like Taylor vortices, but it lacks their rotational symmetry. However, if a spiral flow-pattern is rotated through some angle and then translated vertically by a suitable amount, it ends up looking the same as it started. This is a screw symmetry. When you tighten a nut on a bolt, you're exploiting the screw symmetry of its helical thread: without it, the nut wouldn't be able to turn smoothly and still fit. Spiral flow also has a space–time symmetry. If you look at a spiral flow after some period of time, it still looks spiral, but the spiral is in a diffrent place. The difference can be eliminated either by rotating the spiral until it fits, or by translating it vertically. (These two options exist because the spiral is itself symmetric under a suitable combination of rotations and translations.)

This catalogue can be continued almost indefinitely. All of the 'patterned' states observed in Couette–Taylor flow have their own characteristic symmetries; and all of those symmetries are obtained by breaking some of the space–time symmetries of the apparatus.

Predictive Power

Fine, so the patterns in Couette–Taylor flow can be classified by their symmetries, and appear to be created by symmetry-breaking. So what? Sherlock Holmes has a point to make:

> 'Let me draw your attention to the curious incident of the dog in the night-time.'
>
> 'The dog did nothing in the night-time.'
>
> 'That was the curious incident.'

Not all types of symmetry-breaking actually occur in experiments. There are many mathematical subgroups of the symmetry group of the Couette–Taylor apparatus that don't correspond to observed flow-patterns. A successful theory must account for what doesn't happen as well as for what does; it must explain why forbidden states *are* forbidden. Indeed, it should do much more: it should have some kind of predictive power, as well as rationalizing what's observed after the event. And indeed it's possible to refine our observations about symmetry into a mathematical model which leads not just to qualitative lists of possible symmetries, but to quantitative predictions about which flow-patterns will occur under given conditions.

How does this come about? Well, so far we've ignored an important part of the picture – the laws of physics! We can't expect symmetries to control *everything* about Couette–Taylor flow. Though it's fascinating to see just how much they *do* control: far more than anyone might have expected.

In fact, the physics enters in two distinct stages and on two levels. The first level is relatively coarse: it determines what kinds of instability can occur in the flow-patterns, and provides a short list of 'modes' – which can be thought of as a basic list of patterns out of which all others can be formed by picking suitable combinations. Then the mathematics of symmetry-breaking determines precisely which of these combinations can actually occur, and affords a list of numbers that must be calculated in order to determine the fine details – such as which patterns occur stably. Finally the physical laws come into play again in a far more substantial way, to calculate those numbers. Assembling all of this information together, we obtain precise quantitative predictions that can be – and have been – tested experimentally.

The physical laws that are appropriate to Couette–Taylor flow are the fundamental laws of fluid mechanics, known as the *Navier–Stokes equations*. They were invented by Claude L. M. N. Navier in 1821 and rederived from first principles by Sir George Gabriel Stokes around 1845, and have proved their accuracy over and over again for an amazingly varied range of flows, including 'badly behaved' flows such as breaking waves and turbulence. Despite the eventual success of the Navier–Stokes equations, progress on finding their solutions and on verifying their accuracy proceeded rather slowly. It was not until 1923, when Taylor himself used the Navier–Stokes equations to discover the most basic pattern of all, Taylor vortices, that he also provided the first quantitative test of these equations – a full century after Navier's derivation. With the outer cylinder held fixed, Taylor showed that there is a unique speed of the inner cylinder at which the fully symmetric Couette flow becomes unstable. You get a steady flow with the layered structure of Taylor vortices. So that's one mode: Taylor vortices. The other important mode, studied by Richard DiPrima and his students at the Rensselaer Polytechnic Institute in the 1960s and 1970s, is the one that is observed when the outer cylinder is rotated in the opposite direction to the inner one. Now the initial pattern is not Taylor vortices, but spirals. That's the second mode.

Which pattern occurs – or more precisely, which one occurs stably – depends upon the speed of the outer cylinder. There's a critical speed. If the outer cylinder is counterrotated at less than this critical speed, then the first pattern is Taylor vortices. Otherwise, the first pattern is spirals. If the speed of counterrotation is exactly equal to the critical speed, then both patterns occur simultaneously, and compete with each other to determine the actual pattern. We say that a *mode interaction* occurs at the critical speed. Although you can't see it in the physics, both modes are there all the time. But on one side of the critical speed, the spirals are stable and vortices unstable; on the other vortices are stable and spirals unstable. When spirals are stable, the mathematical model predicts a second bifurcation to Taylor vortices – but this state bifurcates from the now unstable Couette flow! So you won't observe it in the physical system. Similarly, when vortices are stable, there is a second bifurcation to spirals – but again starting from an unstable state. In the mathematical model, the two types of bifurcation switch orders at the critical speed. In the physical system, one disappears and the other suddenly shows up instead.

Once these basic facts have been established, the mathematics of symmetry-breaking makes it possible to write down a general model of behaviour at such a mode interaction, without further reference to the laws of physics. The model has a great deal of freedom, with many adjustable parameters, but in other respects it's highly restrictive. For example the number of variables needed to specify a Taylor vortex state is two (one amplitude and one phase, if you wish). The number needed to specify a spiral state is also two, but spirals can be either left- or right-handed. States in the spiral 'mode' are superpositions of left- or right-handed spirals and thus require four variables. the entire mode interaction thus requires six variables to determine the possible states: accordingly it is called the *six-dimensional model*.

The general analysis of the six-dimensional model started with DiPrima and forms the main bulk of the work that Pascal Chossat, Yves Demay, Gérard Iooss, William Langford and others carried out in recent years. It shows that as well as the component modes – spirals and Taylor vortices – a number of other patterns may occur. These include wavy vortices, twisted vortices, wavy spirals, and a curious state called *ribbons* (Figure 5.4). Ribbons are curious because – at least until very recently – they'd never been observed. There are good reasons for this: if the apparatus has the usual dimensions, then the ribbons state is unstable. Demay and Iooss produced calculations suggesting that ribbons might occur stably if the apparatus was made to different specifications. Randy Tagg has now found strong evidence of the existence of ribbons.

The general mathematics of symmetry-breaking reveals the presence of 'universals' – model-independent phenomena, general kinds of behaviour that occur whenever certain symmetries are present. Such a universal happens here. There's a precise mathematical correspondence between spirals and ribbons on the one hand, and the standing wave/rotating wave pair of states that we've already met in Hopf bifurcation with circular symmetry – the hosepipe example. However, in Couette–Taylor flow the appropriate 'circular' symmetries are vertical translations, together with reflection in a horizontal plane. Taking this difference in interpretation into account, spirals correspond precisely to rotating waves, and ribbons to standing waves. We already know that there's a 'selection rule' for the stabilities of the standing and rotating waves: if one is stable then the

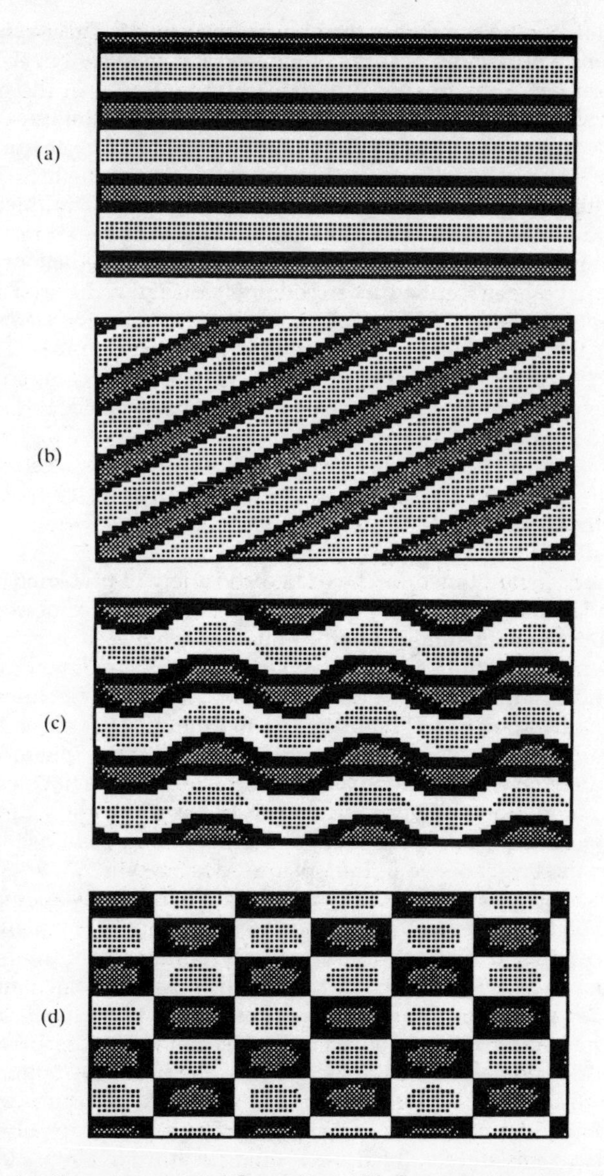

Figure 5.4 Computer-generated flow-patterns. Black represents regions where the axial flow is upwards, white downwards. (a) Taylor vortices. (b) Spirals. (c) Wavy vortices. (d) Ribbons

other is not. So the same goes for ribbons and spirals. This explains the instability of ribbons near the mode interaction: since spirals are observed to be stable, ribbons can't be!

The third stage of the programme, carried out by Langford and collaborators, is to use the Navier–Stokes equations to calculate the precise numerical coefficients that actually occur in the Taylor–Couette system. Then the existence and stability of various states can be read off immediately from the general symmetry-breaking theory. This approach has led to a number of specific predictions, all of which have been verified by experiments of Randy Tagg. These experiments, however, also contain a word of warning for symmetry theorists. For, in addition to confirming the various predictions, Tagg's experiments show the existence of a dynamically complicated state that symmetry arguments seem not to predict; indeed, this state seems to occur only in a finite cylinder.

Apple Turnover

The methods just outlined can be used on a variety of problems, and in particular they shed a lot of light on the general problem of pattern-formation in fluids. Another much-studied area is convection, which we mentioned in chapter 1 in connection with hexagonal patterns of rock mounds. The concept of convection was introduced in the 1790s by Benjamin Thompson, Count Rumford, to explain how heat is transported in . . . an apple pie. Domestic science, indeed.

Around 1900 Henri Bénard investigated convection in a flat layer of fluid, a system that now bears his name. As we said earlier, Bénard convection occurs when a flat layer of fluid is heated from below. If we model the layer by two infinite planes, close together, and apply the same heat everywhere, then there's a tremendous amount of symmetry in the system: all rigid motions of the plane. The only fluid flow with all rigid motions as symmetries is one that's the same throughout the plane: all the fluid must either move straight up, or straight down. It can't do either of those things, because then the entire layer would have to move, so it has to stay still. However, that's not a very stable situation. If the temperature at the bottom of the layer is high enough, there's a strong upward pressure everywhere. If at some point a tiny disturbance lets a few drops of fluid start to rise, then more will follow. If fluid flows up in some places, then it must flow down in others to compensate. So the fluid begins to circulate, and that's convection. Before convection occurs, the fluid stays fixed, but *heat* can still move through it by conduction. After convection starts, the fluid divides into cells, which turn over and over.

In 1916 Lord Rayleigh tried to work out in more detail just what happens when convection starts, and he found that the most basic element of convection is the 'roll' pattern (Figure 5.5). Here the convection cells are roughly cylindrical, and arranged side by side, like planks in a floor. The complete symmetry under all rigid motions has broken: now the pattern is periodic in one direction, across the rolls, but still has complete translational and reflectional symmetry along the rolls. Rolls are mathematically analogous to Taylor vortices, which are really just rolls bent round into a circle.

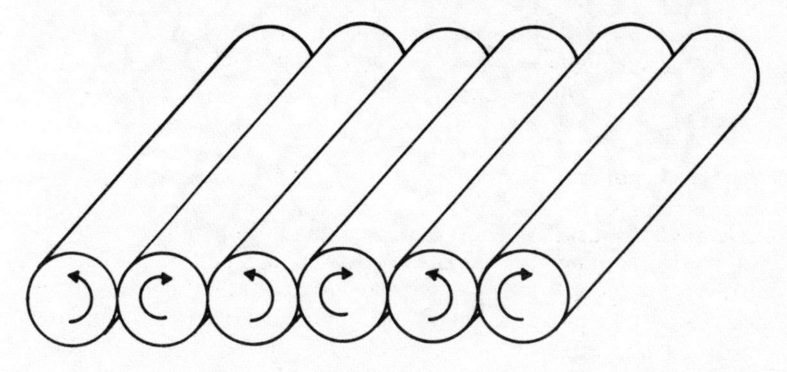

Figure 5.5 The roll pattern in Bénard convection

Combinations of rolls at various angles provide other patterned states, of which the most celebrated is a hexagonal lattice, a mathematical idealization of the polygonal (mostly but not uniformly hexagonal) cells originally seen in Bénard's experiments (Figure 5.6). By using symmetry-breaking methods, it can be shown that there are four basic patterns: rolls, hexagons, triangles, and the patchwork quilt (Figure 5.7).

Bénard convection is going on all around you: something very like it occurs in the weather. Heat from the Sun warms the ground, and the rising hot air forms convection cells. Since the ground is seldom flat, you don't get rolls or hexagons: a more accurate picture is relatively isolated packets of hot air rising like the bubbles in a glass of champagne. These give rise to the 'thermals' favoured by glider pilots as a source of lift. Clouds are one consequence of atmospheric convection, as A. Austin Miller and M. Parry explain in *Everday Meteorology*:

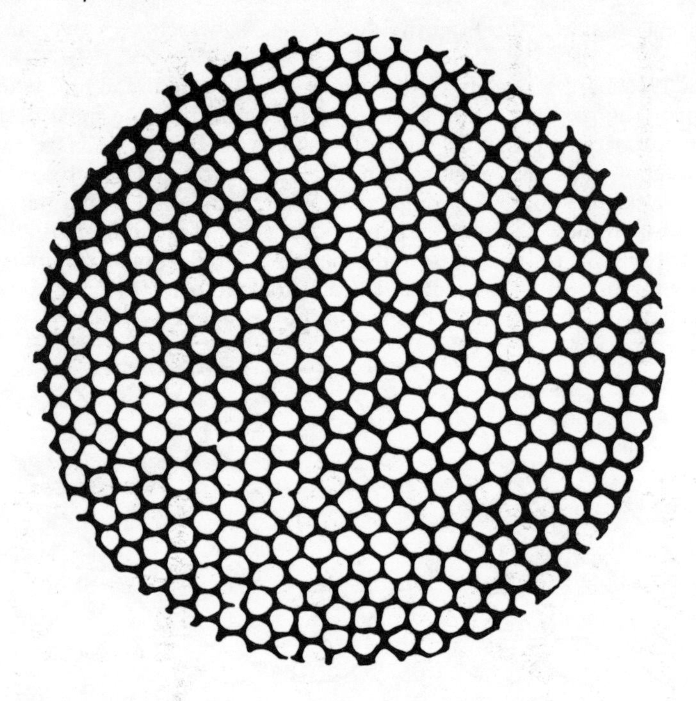

Figure 5.6 Polygonal cells in experimental Bénard convection

> When low-level instability allows convection currents which rise above the condensation level, each is capped by a cloud of the familiar *cumulus* variety. These are often beautiful, sometimes menacing, clouds with flat bases and cauliflower tops. They show up hard and white when lit directly by the sun, dark and with the proverbial silver lining when the sun is behind them. ... The pattern of clouds in a cumulus sky renders visible the initial pattern of thermals rising from the surface, which again depends on the distribution of heat sources. Sometimes the cloud groupings seem quite haphazard, at others it is surprisingly regular. On occasions the cumulus are neatly arranged in rows – *cloud streets* – and this sometimes signifies the drifting of successive thermals downwind from the same source.

A regular stream of bubbles of warm air arising from an isolated 'hot spot' has its own symmetries: spatially it has circular symmetry, temporally it's periodic. Symmetry-breaking is visible in as common a phenomenon as a cloud.

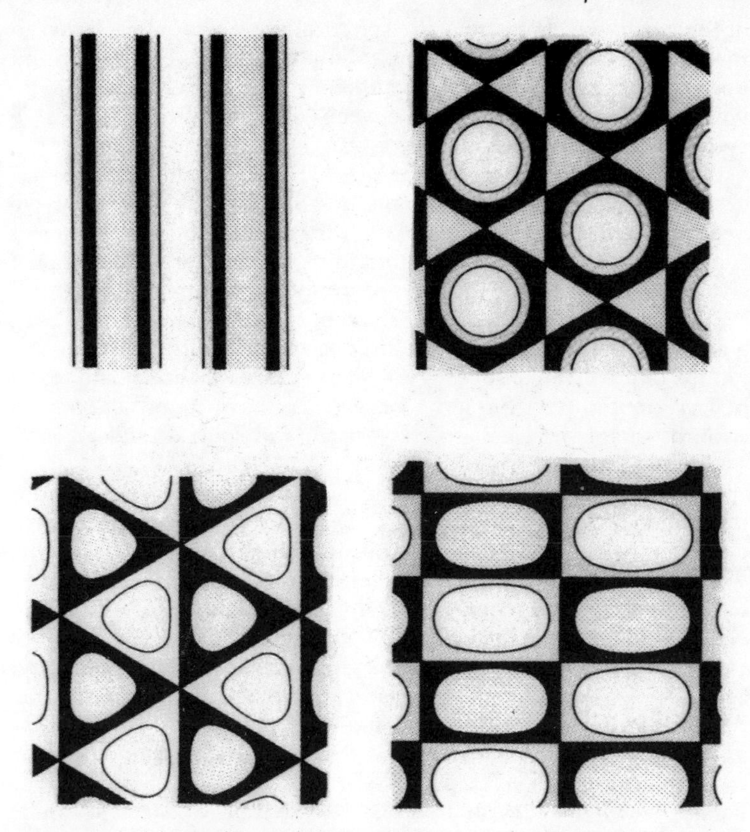

Figure 5.7 Theoretical patterns in Bénard convection. Fluid moves upward in the dark regions and downward in the light regions

Bénard convection is also going on beneath your feet, but you're unlikely to have noticed it. The Earth is a bit like an onion – it comes in layers. Broadly speaking, the continents and oceans float on a solid outer layer, the lithosphere, which in practice is broken into a number of huge pieces, known as plates. Below this is the mantle, made from liquid rock, and inside that is a spherical iron core. The deeper you go, the hotter it gets, so the liquid mantle is heated from below. Therefore it should undergo some kind of Bénard-like convection. However, the symmetry is different from convection in the plane. The Earth is flattened a little at the poles, but only by a small amount, and for many purposes it can be modelled by a sphere. The rotation of the Earth, which destroys the spherical symmetry, is neglected in many geological models, because liquid rock is very sticky and doesn't react rapidly to the forces that a rotating Earth produces. So we're looking at convection between two concentric spheres, a system with spherical symmetry.

What broken symmetries occur here? The fully symmetric state has symmetry group O(3), the symmetries of a perfect sphere. So we need to list all the subgroups of O(3) that can be the symmetries of something; that is, the groups of rigid motions in three-dimensional space that fix a given point (the centre of the sphere). A delightful example of the group-theorist's art has provided mathematicians with a complete list: among its most striking members are the symmetry groups of the Platonic solids – the tetrahedral, octahedral, and icosahedral groups. Fritz Busse and coworkers have found all sorts of patterns in spherical Bénard convection, and some do indeed have the same symmetries as regular polyhedra (Figure 5.8).

Convection in the Earth's mantle is thought to be responsible for the phenomenon of continental drift, which broke up the ancient continent of Gondwanaland and, in the fullness of time, led to the world map that hangs on office walls today. Symmetry-based models of convection in the mantle provide useful insights into what might be going on, and are mathematically tractable. However, they're not the full story, because a map of the globe has no obvious symmetry, certainly not something like an icosahedron. Perhaps the original distribution of continents – or should we say 'continent', since it was all in one piece? – was too asymmetric, so spherical symmetry never really got going. However, polyhedral symmetry does show up in the Earth's gravitational field. To a fairly good approximation, it can be thought of as a spherically symmetric field, plus a smaller circularly symmetric one representing flattening at the poles, plus a *tetrahedrally* symmetric one representing the dominant influence of the uneven distribution of the continents. Plato assigned the *cube* to 'earth', and the tetrahedron to the element 'fire' – but even so, it's some kind of vindication of the Platonic philosophy.

As with warm air, 'hot spots' deep in the Earth can give rise to a plume of rising molten rock. As the plates of the lithosphere drift slowly across the top of such a plume, a chain of volcanoes forms, analogous to cloud streets drifting on the wind above a hot spot on the ground. The Hawaiian islands are an example. You probably never realized that they had anything in common with clouds.

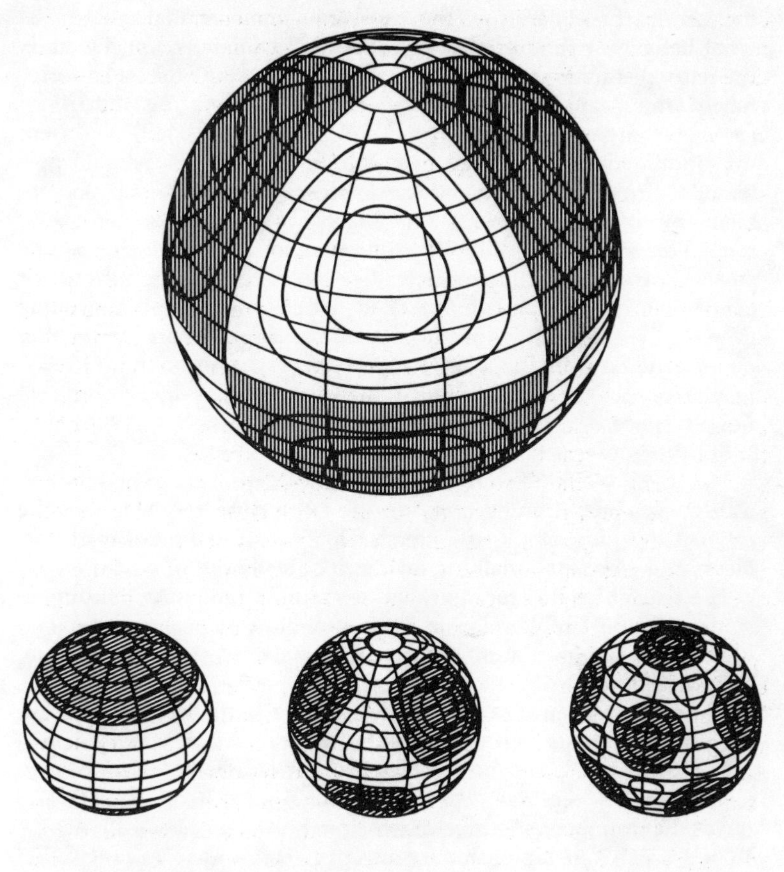

Figure 5.8 Polyhedral Bénard convection in the region between two spheres

New Paradigm

We could carry on for several chapters in this vein, with applications of symmetry-breaking to lasers, flow in porous materials, liquid crystals, elastic buckling, chemical reactions, oscillating molecules, vibrating bubbles – but enough is enough. Time to take stock.

The traditional way to understand what's happening in problems such as Couette–Taylor flow or Bénard convection is to start with careful, precise model equations, based on the physics of fluids and the form of the apparatus. Then those equations are analysed to find what behaviour they predict. In Couette–Taylor flow, for example, the natural equations to use are those of Navier and Stokes. Unfortunately the Navier–Stokes equations can seldom be *solved*, so computer-intensive numerical methods have to be used. And here the traditional methodology runs into a virtually insuperable problem. It's relatively easy, and quick, to solve two-dimensional problems by computer: either two of space, or one of space and one of time. That is, it takes only a few minutes for a computer to handle steady flows in two dimensions. It's a major computation (hours or even days on a supercomputer) to obtain one three-dimensional flow – either a steady flow in three-dimensional space or a time-dependent flow in the plane. It's a colossal computation (weeks, maybe months) to analyse a four-dimensional flow: a time-dependent flow in space.

But that's precisely what wavy vortices are.

Moreover, we don't want just a single flow: we want to understand a three-parameter family (varying the two cylinder speeds and the ratio of the radii of the cylinders). That's a *seven*-dimensional problem, and computationally speaking, it's totally out of reach.

The reason for this rapid growth of running time with the dimensionality of the problem lies in the nature of numerical analysis, and of scientific representation of data. To specify, let alone to calculate, data distributed along a one-dimensional space – a line – we divide the line by a fine grid of points and assign a single number to each. Suppose there are ten grid points – a very coarse grid! Then a one-dimensional problem involves just ten numbers. A two dimensional problem requires a two-dimensional grid, here 10 × 10, and needs 100 numbers. A three-dimensional problem needs 10 × 10 × 10 = 1,000, a four-dimensional problem 10,000, and so on. If we want just to calculate a reasonable representation of wavy vortex flow, we'll need a 100 × 100 × 100 grid in space, and about 1,000 units of time. With each of the three parameters (speed of inner cylinder, speed of outer cylinder, radius ratio) varying through 10 values, we'll need 10^{12}, or one trillion, data points! Suppose it takes ten seconds to solve a one-dimensional problem on a supercomputer. Then it takes 100 seconds to solve a two-dimensional problem; 16 minutes and 40 seconds to solve a three-dimensional problem; and roughly 28 hours to solve the four-dimensional problem. To solve one thousand four-dimensional problems in this manner, required for a survey of how wavy vortex flow depends on the three important parameters, would take about three years!

Those figures are chosen for simplicity, but they're not so far from the truth. Of course, if you *really* need to know the answer, it *can* be

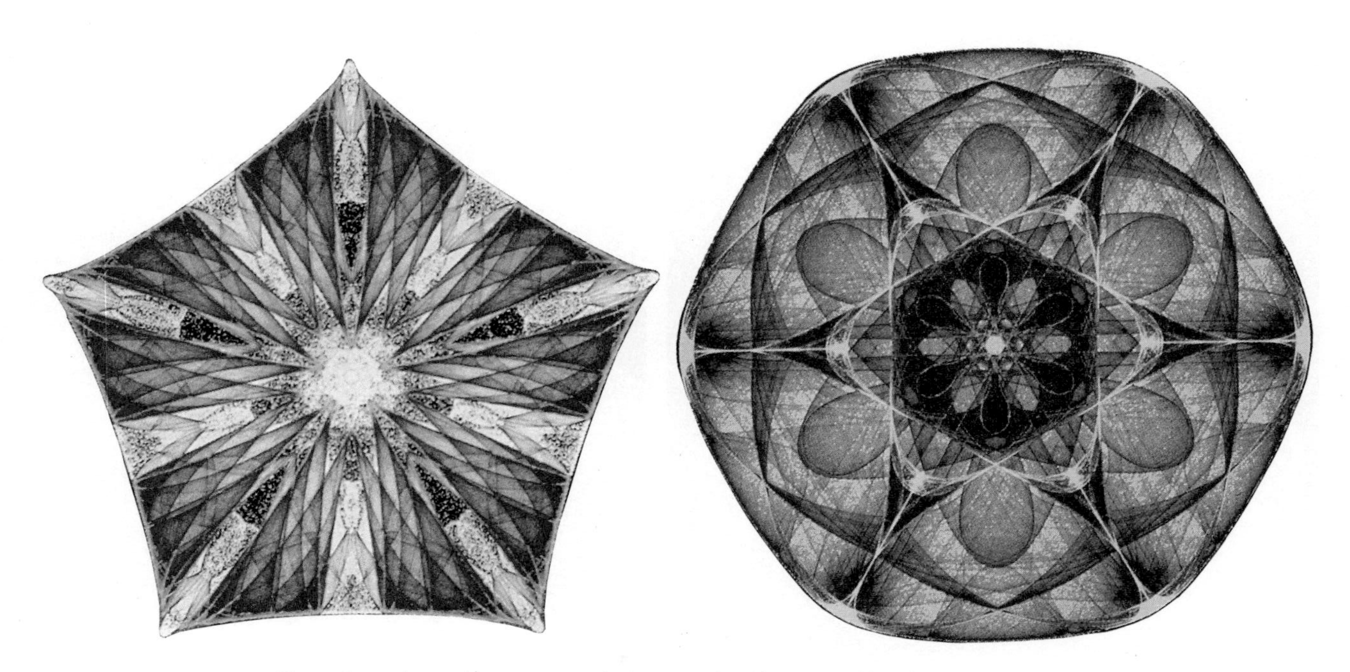

The colour plates show symmetric icons and quilts created by chaotic processes.
See chapter 9 and appendixes 1 and 2.

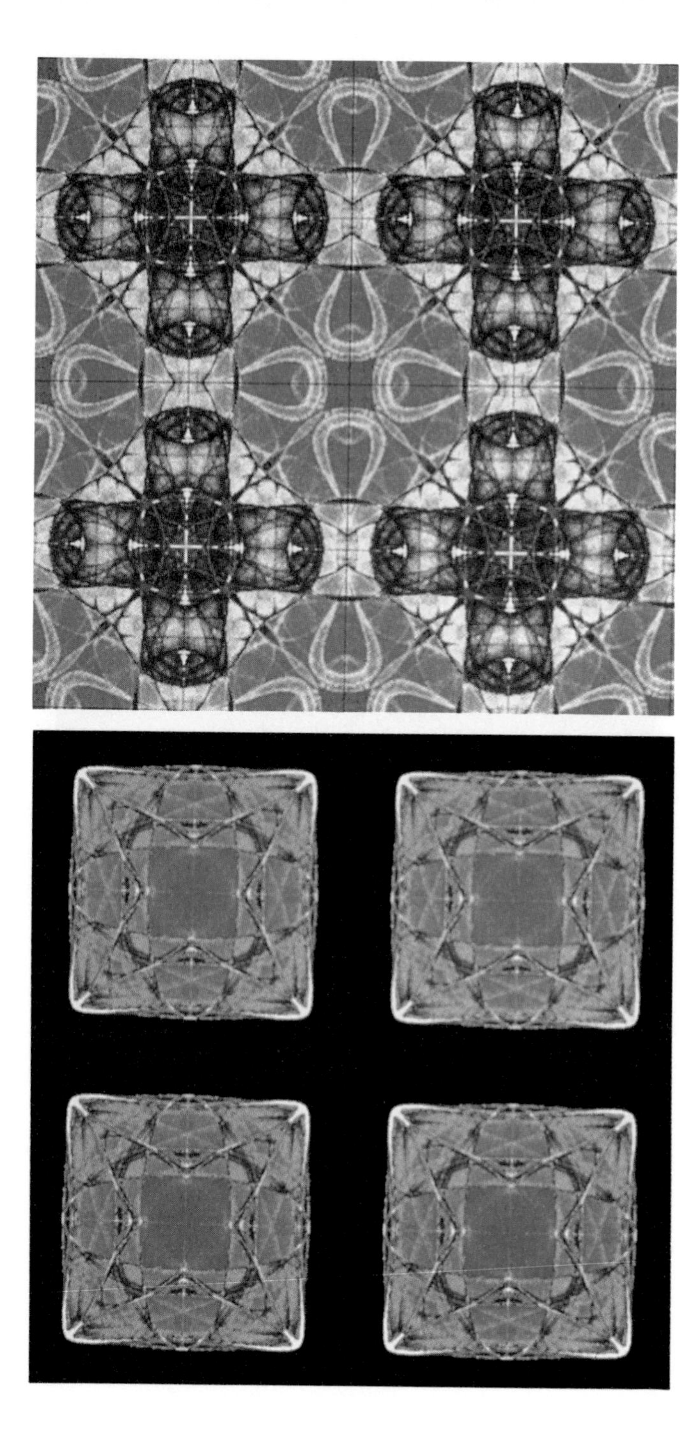

done. However, supercomputers cost around 200 dollars per hour; so the total cost would be around five and a half million dollars. Suppose you've got the five million dollars to spare, and don't mind waiting three years to get the computation done. What, at the end of the day, have you bought? A huge stack of computer pictures telling you 'it does this', but without giving any understanding why. Acceptable, perhaps, if your problem is to check the design of a spacecraft to Mars (who cares why, as long as we know it works?) but useless for gaining scientific insight.

So it would be nice if there were a better way, a new paradigm. Well, there is. The symmetry of the system effectively reduces the dimensionality of the problem. But subtly: if you just assume that the answer must be as symmetric as the problem is, then – rather like the oil company mentioned in chapter 1 – you'll get a wonderfully expensive calculation of a highly unstable, physically impossible, Couette flow; and you won't find Taylor vortices, let alone anything more complicated! Instead, the symmetry has to be fed in at the 'brains' end of the analysis, before using the raw power of the computer. 'Think first, compute later.'

What we must do, therefore, is to begin by asking very general questions, largely unrelated to the actual physics of fluid flow. This goes against the traditional grain in fluid dynamics, but mathematically and scientifically it makes good sense: understand the general nature of the problem first, then use that understanding to plan an attack. It has the advantage of 'technology transfer': the less specific your assumptions about the physics are, the more widely applicable your results may be. So we ask: what are the symmetries of the system? What is the catalogue of possible types of pattern in a system with that symmetry? What must we calculate to determine which of these possible patterns actually occurs, and what its stability is? Only after answering these general questions do we sit down and calculate those numbers. Admittedly we don't get as much detail as we might by a full-blooded numerical simulation: that's one price we must pay. But in compensation we get a broad understanding of the mathematical features not just of his system, but of any other system with the same symmetries. We use model-independent concepts as far as possible, and only put in the detailed physics of the model at the end. Symmetry selects the general range of possible patterns; physics tells us which patterns actually occur.

The best thing of all, of course, is that both approaches are available if you want them. We're not fighting an election, where only one party wins; we're not arguing whether symmetry considerations are inherently superior to numerical analysis or vice versa. What we're saying is that there are alternative strategies available, depending on what you want to know, and that different strategies are effective on different types of problem. The application of mathematics to science is an ongoing process in which we learn what the best tools are for any given job. The generalities of symmetry-breaking provide not just a new tool, but a whole new toolkit.

6

The Universe and Everything

> The Great Green Arkleseizure theory is not widely accepted outside Viltvodle VI and so, the Universe being the puzzling place it is, other explanations are constantly being sought.
>
> Douglas Adams, *The Restaurant at the End of the Universe.*

People have always been fascinated by the heavens: the revolving bowl of stars, the changing phases of the Moon, the rare spectacle of a total eclipse of the Sun, the looping paths of wandering planets threading through the constellations. With the invention of the optical telescope, and its sophisticated successors such as radio, infra-red, and x-ray telescopes, and with the dispatch of electronic probes throughout the Solar System, we have begun to glimpse the true complexity of the cosmos. And for every feature of it that we understand, there are a hundred that we don't. How are comets formed? Why are the gaps in Saturn's rings spaced the way they are? What causes the strange radial 'spokes' in those rings, discovered by the *Voyager* probes? Why are galaxies spiral? Why is the matter in the universe distributed in enormous clumps, rather than being evenly spread? Humanity still invests an incredible amount of time and effort trying to answer such questions, even though they concern events a million miles, or even a million light-years, away.

There's *some* practical spin-off, certainly – but much of that would have come about in some other manner. We live in a wondrous universe, so vast that it can make us feel no more significant than an ant, yet, in contrast, so awe-inspiring that we feel we belong to something of immeasurable importance. The reason for trying to understand the universe isn't that we thereby blunder into a new material for coating non-stick frying-pans. It's that we gain insight into our place in the scheme of things, and of just how wonderful and unexpected that scheme can be. The aim of science is not just the manufacture of new toys: it's the enrichment of the human spirit. This isn't an astronomy book, and many of the celestrial marvels are outside its brief. But in the heavens, too, we find many examples of patterns brought about by broken symmetry. Our unifying principle functions on a cosmic scale as well as a terrestrial one.

What Shape Is a Star?

A star is a vast mass of gas, mostly hydrogen and helium, brought together by its own gravitational attraction. The most symmetric shape for a star would appear to be a sphere; and a body of matter with spherical symmetry might be expected to retain that symmetry as it pursues its dynamical evolution.

Of course, it's not that simple.

One complication is that stars rotate. It took humanity a long time to make this discovery, because stars are so far away that they show no disc – except for one that's much closer to home, our Sun; and that's so bright that it's impossible to observe any detail on its surface using the naked eye. With a telescope, however, and suitable filters to protect the eye, such features can be observed – and so can the way they move. In June 1611 Johannes Goldschmidt made the first public announcement that the surface of the Sun isn't perfect, but has blemishes: sunspots (Figure 6.1). Goldschmidt is more commonly known by his Latin name, Fabricius, and we'll encounter him again in chapter 8 in connection with the gaits of animals. Fabricius and Galileo were rivals, and Galileo promptly stated that he had made similar observations in 1610, although he published nothing until 1612. By then Thomas Harriot, in England, and Christophe Scheiner, in Germany, has also seen spots on the Sun. Scheiner was a Jesuit priest, and had been warned by his superiors against believing in the reality of sunspots, so he published his discovery under a pseudonym. His explanation that the spots were small planets revolving round the Sun was contested by Fabricius and Galileo, who thought they were either on the Sun or very close to it. In 1613 Galileo responded to Scheiner's letters with three of his own, in which he also declared his belief that the Earth goes round the Sun, rather than the Sun round the Earth as was generally believed at the time. It was this that put him on a collision course with the Roman Inquisition and caused so much grief later; but for our purposes the most significant feature of the letters was his realization that the Sun must rotate.

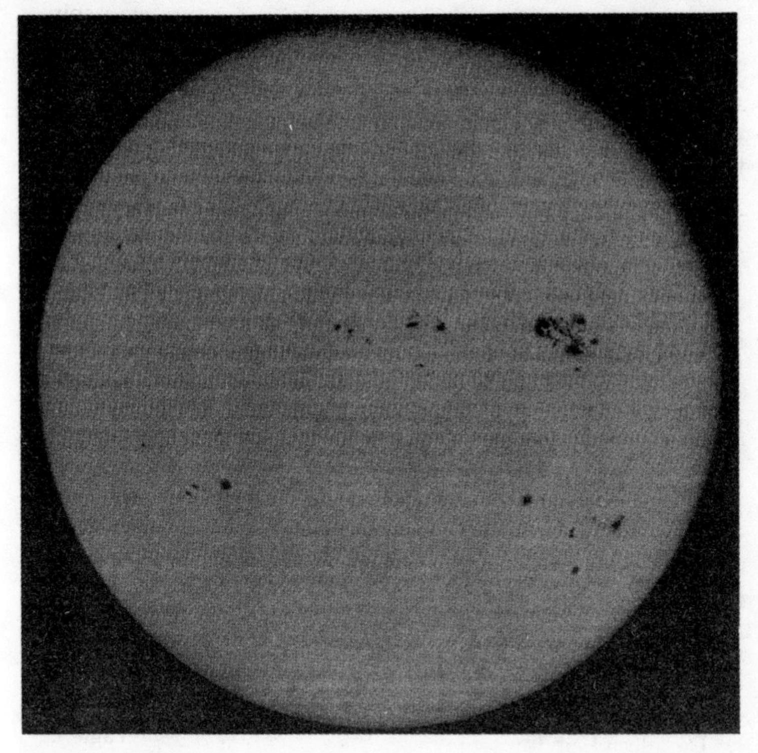

Figure 6.1 Sunspots

Before the discovery of sunspots, with no discernible features on the Sun's disc, there had been nothing that might make any rotation visible; but Galileo noticed that sunspots migrate across the disc of the Sun, always in the same direction and always taking roughly 14 days to cross from one edge to the other. Near the edges, the apparent speed with which they move *decreases*; and Galileo correctly deduced that this is the result of foreshortening, as the spot follows the curve of the Sun. This led him to conclude that the spots must be on or near the surface. Scheiner eventually realized that Galileo was right, and made observations leading to an estimate of about 27 days for the Sun's rotation period. Nothing much was done to further our understanding of the rotation of the Sun for about two hundred years. Then, in the 1850s, Richard Carrington and Gustav Spörer independently discovered that the Sun doesn't rotate like a rigid body. The rotational period depends upon latitude: it's slowest at the equator and increases towards the poles.

What are stars made of? In 1825 Auguste Comte declared that this is a question we'll never be able to answer, and was promptly proved wrong when spectroscopy was invented. In 1666 Isaac Newton had discovered that when sunlight shines through a prism, it breaks up into a band, or *spectrum*, displaying all the colours of the rainbow. Indeed raindrops affect sunlight the same way: the colours literally *are* those of the rainbow. Light of different wavelengths – which the human eye sees as different colours – is bent through different angles by the glass of the prism. In 1802 William Wollaston noticed dark lines in the spectrum of the Sun, as if some wavelengths were missing (Figure 6.2). Experiments showed that if light is passed through a flame, then dark lines appear in the spectrum, where substances in the flame absorb light of a characteristic wavelength. The Sun's spectral lines are caused by substances within it, which absorb the 'missing' wavelengths of light. So, by observing the spectrum of a star, and seeing where the dark bands occur – a technique called spectroscopy – you can find out what it's made of.

You can also find out how fast it's moving, because of an effect discovered by Christian Doppler in 1842. When light waves are emitted from a moving body, the motion changes their effective wavelength. Bodies moving away from us shift their light towards the red, those moving towards us shift it into the blue. When an ambulance whizzes past and its siren seems to change pitch, that's the Doppler effect for sound waves.

Accurate spectroscopic measurements of the rotational speed of the Sun were made in the 1870s by Hermann Vogel, Nils Dunér, and Jakob Halm, using the Doppler effect. In 1877 Sir William de Wiveslie Abney suggested that the Sun's rotation might show up as a broadening of its spectral lines. Vogel disagreed, and the idea found no favour until 1898, when he made a complete U-turn, having himself observed broadened spectral lines in the Sun after introducing photographic methods. This discovery was important because the

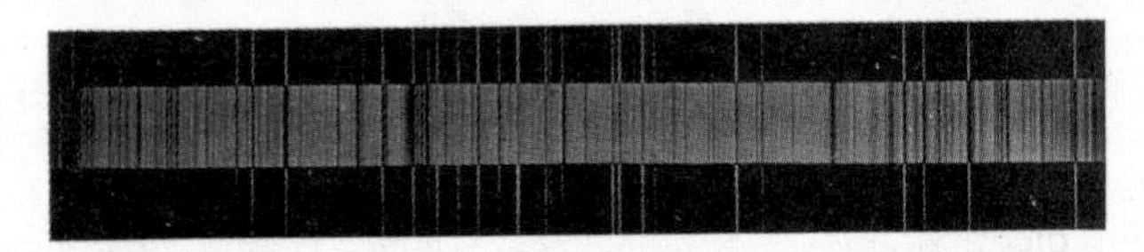

Figure 6.2 Absorption lines in the spectrum of the Sun

spectral lines of distant stars can be observed, whereas their analogues of sunspots – if any exist – cannot, because the stars are too far away to show a disc. By 1909 it was clear that the Sun was not alone: Frank Schlesinger showed that at least one other star, the eclipsing binary delta Lyrae, exhibited broadened spectral lines and therefore must rotate. Soon lambda Tauri, beta Lyrae, and beta Persei joined the list.

The Great Flattener

Meanwhile the mathematicians had been busy. Their interest in the problem of a rotating fluid mass, either liquid or gaseous, goes back to Isaac Newton, who in 1687, in volume 3 of his famous *Mathematical Principles of Natural Philosophy*, considered the shape of the Earth. Newton favoured an oblate spheroid, that is, a sphere slightly flattened at the poles but retaining circular symmetry about its axis. He imagined such a spheroid, made of fluid, and asked how its shape would depend upon its speed of rotation. The spheroid must be in *equilibrium*, with the centrifugal forces caused by rotation exactly balancing the attractive forces of gravity; and this relation links the rotational speed to the shape. Newton proved that the eccentricity of the cross-section – the amount of flattening – must be $\frac{5}{4}$ times the ratio of the centrifugal force at the equator to the average gravitational attraction at the surface.

A rival to Newton's ideas was the vortex theory of René Descartes. According to its adherents, this theory implied that the Earth should be *stretched* at the poles, rather than flattened. In the 1730s Pierre Louis Moreau de Maupertuis led an expedition to Lapland – one of several sent to various parts of the Earth – to measure the length of a meridian. Maupertuis confirmed that the Earth was shaped in accordance with Newtonian theory, and not Cartesian, thereby gaining the nickname 'the great flattener'. That was the end of vortex theory.

In 1737 Alexis-Claude Clairaut worked out the formula for the gravitational attraction of a slightly flattened sphere; this was generalized to arbitrary spheroids by Colin Maclaurin in 1740. He proved that any oblate spheroid is a possible equilibrium shape for a rotating fluid mass. These *Maclaurin spheroids* are axially symmetric, but not spherically symmetric. Although nobody at the time phrased it this way, Maclaurin had demonstrated that the rotation of a spherical body causes its symmetry to break, from spherical to circular: see the second picture in Figure 6.3. This really is no great surprise, because when rotation is present the original spherical symmetry is replaced by circular symmetry about the axis of rotation: the type of symmetry-breaking occurring here is not spontaneous, but induced: the system as a whole changes symmetry, and not just its states. However, roughly a century later – in 1934 to be precise – Karl Jacobi found a true example of spontaneous symmetry-breaking in the shape of a rotating fluid mass. He provided strong evidence for the existence of non-axially symmetric ellipsoidal solutions: ellipsoids with all three axes unequal. Joseph Liouville completed the proof in the same year.

The form assumed by a rotating mass of fluid depends upon its angular momentum – how much matter is rotating and how fast. For Maclaurin ellipsoids, the shape varies from a perfect sphere to a flat disc as the angular momentum increases from zero to infinity. The Jacobi ellipsoids vary from axially symmetric spheroids to infinitely long needles, but they can occur only if the angular momentum exceeds a particular value. This is the value at which the axial symmetry is broken. Jacobi ellipsoids also have some symmetry, namely, they're unchanged by rotation through 180°, and also by top–bottom reflection. In 1885 Henri Poincaré showed that a further breaking of the rotational symmetry can occur: pear-shaped figures can branch off, or (as he called it) *bifurcate*, from the Jacobi ellipsoids. Indeed it was this discovery that initiated the area now known as bifurcation theory. The Russian mathematician Aleksandr Mikhailo-

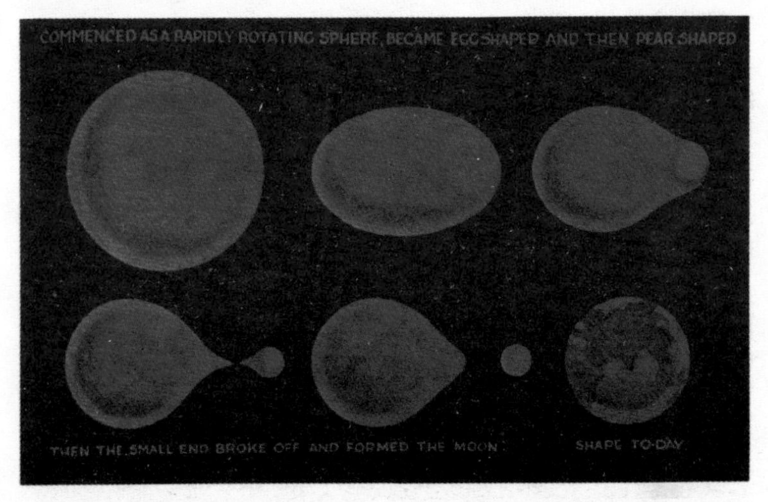

Figure 6.3 The pear-shaped blob theory of the formation of the Moon. The second stage is a Maclaurin spheroid

vich Liapunov had made the same discovery a year earlier, but his results were largely unknown in the West until Poincaré asked for them to be translated. This language problem still bedevils East-West scientific relations and causes all kinds of priority disputes. The pear-shaped figures found by Liapunov and Poincaré aroused tremendous interest, because it was thought that as the angular momentum increased, the pear might break up into two blobs (Figure 6.3). Lord Kelvin and Peter Guthrie Tait suggested this as a mechanism for the creation of double stars; it was also proposed as a theory of the formation of the Moon.

Island Universes

If you look up on a clear night you will see hundreds of bright stars, scattered randomly across the sky. Sometimes several appear to be grouped close together, but on the whole they are fairly equally distributed. Across the night sky there also runs a huge, irregular, luminous band. On a summer evening it rises from the north east, passing through the constellations of Persues, Cassiopeia and Cepheus. Its brightness increases substantially as it passes Cygnus, where it forks into two unequal tracks, and it disappears below the south-west horizon through Scorpio and Sagittarius. The huge, irregular luminous band is the Milky Way (Figure 6.4).

Figure 6.4 Our galaxy, the Milky Way

Though mysterious to the naked eye, the Milky Way reveals its true nature to any small telescope. It is composed of stars. Lots of stars, probably around a hundred billion. The first study of the distribution of stars in and around the Milky Way was made around the end of the eighteenth century by William Herschel, by the simple if tedious expedient of counting how many stars he could see through his telescope in various regions. He discovered that the apparent density of stars is greatest along the Milky Way, and decreases steadily away from it. Moreover, this decrease is more noticeable for faint stars than for bright ones. The brightness of stars is expressed by a number called their *magnitude*: this ranges from 1 for the brightest stars through 5 for the limits of visibility down to 20 or so for what can be seen in a good optical telescope. At magnitude 20 there are roughly fifty times as many stars per unit area near the Milky Way as there are at those parts of the sky that are farthest away from it.

In the mid-eighteenth century the German philosopher Immanuel Kant and the Englishman Thomas Wright suggested that the system of stars forms something like a flattened disc, with the Sun being inside; and Herschel's observations confirm this idea. The argument is simple. If our own Sun, and with it the Earth from which we observe the heavens, is buried within a disc of stars, then more stars will be seen in directions parallel to the disc than at right angles to it. The effect is more noticeable for faint stars because the brightest stars tend to be the nearest ones, and the effect of flattening only becomes apparent when we observe our surroundings at distances that exceed the thickness of the disc.

Herschel at first thought that the Sun lies at the centre of this disc, but as always, humanity's egocentricity proved misplaced. The evidence is plain to the naked eye: the greater brightness of the Milky Way in the direction of Cygnus can most naturally be explained if we assume that the Sun lies away from the centre of the disc, so that there are more stars on one side of it than on the other. Thus the centre of the disc presumably lies at some considerable distance from the Sun, in the general direction of Cygnus.

This hypothetical disc of stars was humanity's first inkling of the existence of the structure now known as a *galaxy*. It soon became clear that the galaxy in which we live is not alone. Indeed, long before Kant's time, astronomers had been intrigued by faint, fuzzy patches of light that they sometimes saw through their telescopes when they expected to see the bright pinpoint of a star. They called them *nebulae*. Kant not only suggested that the Milky Way is a disc of stars: he also suggested that each nebula is a similar disc, immensely distant. Herschel's confirmation of Kant's theory of the Milky Way led to an acceptance of the whole idea, that the universe is composed of innumerable galaxies, each galaxy being composed of innumerable stars. It was a picture with great appeal to a sea-faring age: Alexander von Humboldt called the galaxies 'island universes'.

The next important step towards the understanding of galaxies was taken by the Frenchman Charles Messier who, paradoxically, wasn't really interested in galaxies at all. He wanted to study comets. But all too often, when he observed what he thought was a new comet, it turned out to be a galaxy. This was very annoying, and Messier built up a catalogue of galaxies in order to avoid making the same mistake again. When he published it in 1784, he had found exactly 103 galaxies, which astronomers referred to by his numbering system, as M1, M2, ... and so on, all the way to M103.

These galaxies were so far away that the telescopes of the day were unable to resolve them into their component stars. Although it was widely believed that once telescopes powerful enough were built, the expected stars would duly materialize, nobody had any serious evidence that this belief was correct. When the evidence duly arrived, it wasn't that simple, of course. Spectroscopic methods showed that some nebulae are composed not of stars, but of gas! However, by 1920 Edwin Hubble, using the 100-inch Mount Wilson telescope, had partially resolved the stars in the nebula M31 – number 31 in Messier's catalogue – in the constellation of Andromeda. He also managed to work out roughly how far away M31 is, by exploiting the behaviour of a class of special stars known as *cepheids*.

Pulsating Stars

If you look down from an aircraft at night, flying over an urban area, you see lights. Some are dim, some bright. *How* bright depends upon two things: how much light they're emitting, and how high up the plane is. If you can't recognize the nature of individual lights then you can't use them to work out the height of the aircraft: they might be bright lights a long way off, or faint lights much closer. However, if you can tell that some of them are sodium-arc street lights, say, then you're immediately provided with a yardstick. The true light output of a sodium-arc streetlight is a known constant value. By comparing the observed brightness with this standard value, you can easily work out the distance.

In astronomy, cepheids play the role of sodium-arc streetlights. The perceived brightness of a star depends upon the same two factors: its true or intrinsic brightness, and its distance. If you see a faint star, it could either be a bright star a long way off, or a fainter one nearby. But how can we tell which? Cepheids enter the picture because the light that they emit isn't constant: instead, they're variable stars, whose light output pulsates. Moreover – and crucially – the period of the pulsation is related to the true brilliance of the star; so by observing the period, you can deduce the true brightness. That is, cepheids function not so much as a single standard type of streetlight, but as a whole family of streetlights of different – but known – luminosities. This is only useful if you can tell that a given star *is* a cepheid, without knowing how far away it is; and by good fortune it turns out that you can recognize a cepheid by its spectrum, independently of its distance or true brightness. So, if you see a cepheid pulsating, you can measure its period, deduce its true brightness, compare with the brightness you've observed, and work out how far away it is.

There were cepheids in M31 – the first was found in 1923 – and they turned out to be roughly a *million* light years distant. Kant was right, but the distances were far greater than he or anyone else had imagined.

Whirlpools of Light

The galaxies weren't just discs, either. Many had internal structure. The most dramatic were magnificent whirlpools of light, multiarmed spirals, like a Catherine-wheel. M31 is an example, although it is tilted away from the line of sight and flattened into an ellipse by perspective. The galaxy M51 in the constellation Canes Venatici (Figure 6.5) is a spectacular circular spiral with a secondary concentration of stars like a blob attached to one of its spiral arms, so that it looks rather like a comma.

Spirals arms aren't the only possible internal structure. Hubble classified galaxies into four main types:

- *Elliptical galaxies* Smooth, featureless blobs containing little gas or dust, generally lacking a sharp outer edge. There is no strong consistent rotation: individual stars seem to be moving in random directions.
- *Lenticular galaxies* These have a prominent disc, containing no gas, no bright young stars, and no spirals. They are smooth and featureless, like elliptical galaxies, but the distribution of brightness is quite different, being closer to that of spirals.

Figure 6.5 The Dramatic spirals of M51

- *Spiral galaxies* These have a prominent disc, including gas and dust as well as stars. There are spiral arms, of somewhat variable geometry; some are tightly wound while others are long and thin. The whole disc rotates, with the speed of rotation being roughly independent of the distance from the centre. (Incidentally, the motion is not visible: it takes the Sun about two hundred million years to complete one rotation of the Milky Way. But spectroscopic measurements can give clues as to how fast and in what direction the motion is.)
- *Irregular galaxies* As the astronomers James Binney and Scott Tremaine say: 'any classification has to contain an attic'. Galaxies that aren't of the three previous types are consigned to this fourth classification. Some may be spiral or elliptical galaxies that have been distorted by a close encounter with a neighbouring galaxy. Others, like our nearest neighbours the Magellanic clouds, are just low-luminosity gas-rich systems. These do not have the same well-defined circular motion of lenticular and spiral galaxies; but they don't exhibit the

random motion of elliptical galaxies, either. Their motion is structured, but more swirling, rather like turbulent flow of a fluid.

The four main types of galaxies merge continuously into a single sequence, the *Hubble sequence*:

elliptical → lenticular → spiral → irregular.

For example the proportion of gas to stars increases as we go along the sequence, and the pattern of rotation changes in a consistent fashion. Actually, Hubble distinguished two types of spiral galaxy: the *normal spiral*, and the *barred spiral*, which has a clearly defined straight bar across its centre. So Hubble's picture of his sequence (Figure 6.6) splits into two branches – perhaps another example of bifurcation.

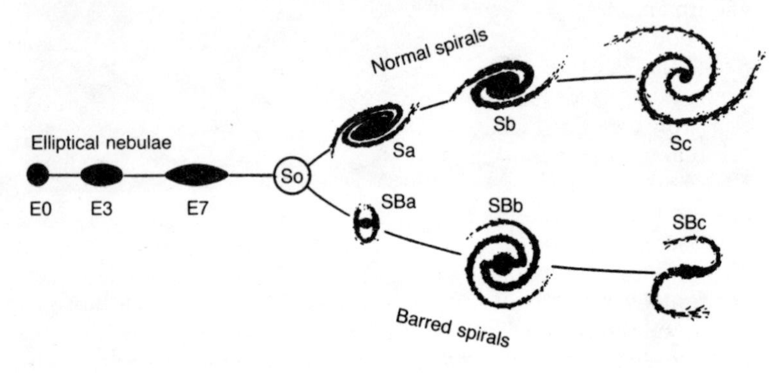

Figure 6.6 Hubble's picture of his sequence

Why Spirals?

'Spiral structure,' say Binney and Tremaine, 'has proven to be one of the more obstinate problems in astrophysics.' Until the early 1960s, virtually all astronomers thought that the spiral arms are caused by the interstellar magnetic field. Only one, the Swede Bertil Lindblad, had the right idea: that spirals are purely dynamic in origin, caused by the gravitational interaction of the stars in the galaxy. He worked on the problem, without great success, from 1927.

Just before Lindblad's death in 1965, C. C. Lin and Frank Shu had the crucial insight that crystallized Lindblad's ideas into a well-formed theory. Up till then, there had been a tendency to assume that the spiral arms were fixed features, in the sense that a particular star is either in a spiral arm, or not, and as time passes, it stays there. As the galaxy revolves, the arms revolve with it, and so do the stars in the arms. This belief poses enormous problems if you think that only gravitational interactions are relevant: effectively the entire spiral structure has to rotate 'rigidly'. But then centrifugal force increases as you move away from the centre, while the gravitational attraction decreases; so the forces are unable to balance. A rigidly rotating galaxy will tear itself apart.

Lin and Shu realized that there's another possibility: the spiral structures might be density waves, sweeping through the system of stars. Such a wave can maintain its overall coherence and structure, even if the stars move into and out of the regions of greatest density – the arms. This is easy to understand: we see examples every day. Take a length of rope, tie it to a wall, and move the other end up and down, sharply. A wave travels along the rope from your arm, bounces off the wall, and returns. But do *individual bits of rope* travel along to the wall and back? Of course not! A slightly more subtle example: drop a stone in a pond, and watch the circular ripples that grow. They *seem* to move outwards, but does the water actually move any great distance? It doesn't, as you can check by watching what happens to leaves or sticks floating on the surface as the wave passes. The water at any given place just bobs up and down.

These aren't waves of density, mind you, so it might be wise to offer two further examples. Instead of waving a rope, imagine a long horizontal spring, a soft, sloppy one like a 'slinky' toy, with biggish, closely spaced coils. Fix the far end to a wall and give a sharp push at the free end. You'll see a compression wave travelling along the spring: the coils bunch up and the bunch travels along. But do bits of the spring travel with the bunch? They can't possibly, or the whole spring would end up flattened to the wall. What happens is that the front of the bunch hits a new coil, picking it up, but it also drops one off the back. So the closely-packed bunch of coils travels along without losing its identity, but no individual coil stays in it for more than a short time, and on average each coil ends up back where it started. That's a typical density wave.

The final example is sound. Sound is a density wave transmitted through the atmosphere. But the individual molecules of air don't travel along with the sound: they just help the wave on its way as it comes past them, by jiggling a bit when it arrives and nudging the next molecule in the right direction. If the air went along with the sound, then every time you spoke to somebody they'd be standing in a howling gale, emanating from your mouth.

Density waves alone aren't enough to solve the problem of spiral structure, but Lin and Shu went further. They assumed that the spiral density wave isn't a short-lived phenomenon, but a kind of steady state behaviour, which we now call a relative equilibrium. (Astronomers call it a quasi-stationary state.) If you ignore the rotation of the galaxy, the pattern of spirals should always look pretty much the same – unless you work with such a long timescale that the actual composition of the galaxy is changing as it ages.

The Lin-Shu hypothesis, that spirals' arms are a density wave in relative equilibrium, is a godsend to theorists, because it makes all sorts of calculations and predictions feasible. The reason is that, apart from the rotation, the density pattern remains constant in time; and constants are a lot easier to calculate than variables. Adding in the rotation is easy. The Lin-Shu hypothesis explains a lot of the known features of spiral galaxies, but it's still the subject of intense debate.

It's also such a simple theory that it looks almost obvious. Why did it take astronomers so long to see what might be happening? Perhaps they knew *too much* about galaxies, and too little about the mathematics of symmetry-breaking. For example, the way the observed rotation rates of galaxies changes with the distance from the centre is inconsistent with the Lin-Shu hypothesis. Demise of hypothesis? No, it's not that simple, because the same data are inconsistent with the laws of gravitation. *Unless* there is a great deal more matter in a galaxy than we can see through telescopes – but then we can resuscitate the Lin-Shu hypothesis! The 'missing' matter is called *cold dark matter*, and we will encounter it again at still larger cosmic scales later on. Another objection is that near-collisions between elliptical galaxies can create spiral arms, dragged out by gravity much as the Moon causes tides on Earth, but on a far vaster scale; so spirals need not be natural consequences of the dynamics of a *single* galaxy, as Lin and Shu maintain.

With a large dose of twenty-twenty hindsight, we can (and will) re-interpret the Lin-Shu theory as an example of symmetry-breaking, and see that the mechanism they propose, while not perhaps inevitable, is extremely natural and plausible. The crucial observation from this point of view is that a rotating spiral structure has quite a lot of symmetry, at least in an idealized setting. Because the generalities of symmetry-breaking are model-independent, neither cold dark matter nor near-collisions cause problems to the abstract line of argument – though they certainly affect its detailed working out!

The relevance of symmetry-breaking is very clear in some of the earliest numerical experiments, carried out in 1971 by F. Hohl. The computation does *not* build in the Lin-Shu hypothesis as an assump-

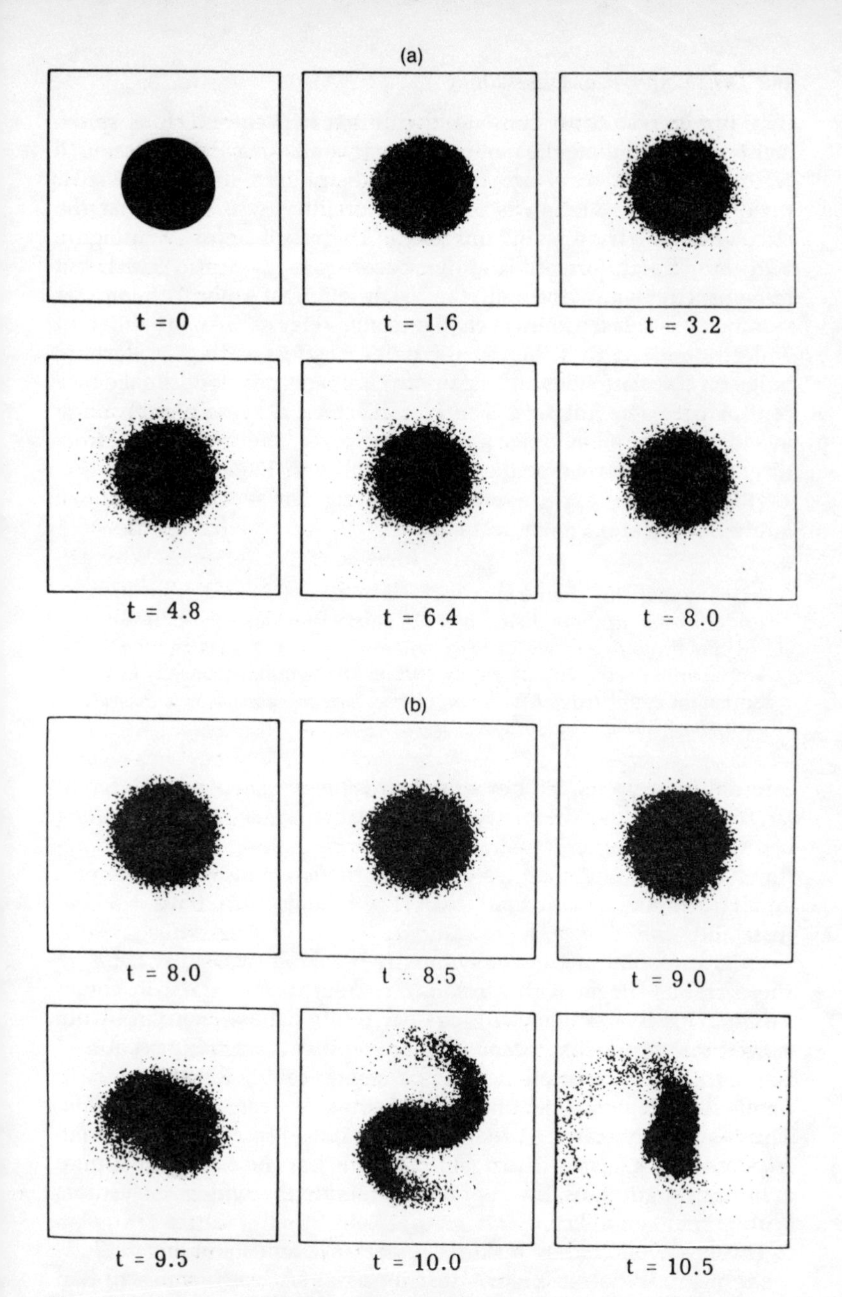

Figure 6.7 Computer simulation of the dynamics of a circular disc of stars. (a) Requiring symmetry to be preserved. (b) Allowing it to break

tion, but its results are consistent with that hypothesis. Hohl used a computer to simulate the motion of a disc containing 100,000 stars. To begin with, the stars are uniformly distributed inside a circular region; and the motion is simulated on the assumption that the circular symmetry remains unbroken. The results, shown in Figure 6.7a, are straightforward: the disc becomes fuzzy at the edges, but remains circular in form. If the assumption of unbroken circular symmetry is relaxed, however, the results (Figure 6.7b) are different and striking. At first the disc becomes fuzzy; but then it starts to bulge on opposite sides and a pair of spiral arms develops, linked by a central bar. The rotating disc is subject to a symmetry-breaking instability. Not all symmetry is lost, however: the two-armed structure has a single rotational symmetry, through 180°.

This numerical experiment is fascinating, for several reasons. To quote Binney and Tremaine again:

> In many cases the shapes of spiral galaxies are approximately invariant under a rotation about their centers. A galaxy that looks identical after a rotation through an angle of $2\pi/m$ radians [that is, $360°/m$] is said to have m-fold spiral symmetry. A galaxy with m-fold symmetry usually has m dominant spiral arms. *Most spiral galaxies have two arms and approximate twofold symmetry.*

The italics are ours. Binney and Tremaine are saying, in different language, that the symmetry group of such a galaxy is Z_m. Strong central bars are often apparent in real spiral galaxies. Indeed sometimes the bar is the main feature, and arms are almost non-existent. Analytic models discovered by A. J. Kalnajs also have the bar instability.

From the point of view of symmetry-breaking, we can summarize these results. Begin with a rotating circular disc of stars, in equilibrium. This is a system with circular rotational symmetry (but not reflectional symmetry, because this would reverse the direction of rotation). If the variation of some parameter causes the symmetry to break, then *the only possibility* is m-fold symmetry for some value of m. Thus we can 'postdict' – predict after the fact – the observed symmetries of spiral galaxies. The detailed form is beyond the reach of pure symmetry arguments, but m-armed spirals are the simplest structures with m-fold symmetry.

This is a good place to make an important point about the methods we're using. What we've just described is a *mathematical* description of how we might understand the existence of solutions to the dynamical equations that possess spiral symmetry. The parameter that causes symmetry to break need not be physically realized in the actual evolution of the galaxy. It's not necessary for the galaxy itself to begin with circular symmetry and then lose it. It's sufficient that the abstract mathematical space of all possible solutions to the galaxy equations contains paths along which such changes occur, because that provides a mathematical rationale for the *existence* of spiral solutions. Once we know they exist, we know they can occur in nature. Having made that point, we're also not saying that the galaxy *doesn't* evolve from circular to spiral symmetry. As Hubble said of his sequence: 'The terms "early" and "late" are used to denote relative position in the empirical sequence without regard to their temporal implications.' Symmetry-breaking classifies what's possible: it doesn't tell you what actually occurs. For that, more detailed dynamical theories are required. One recent theory starts with irregular galaxies and runs *backwards* through the Hubble sequence.

The predominance of two-fold symmetry can't be explained on pure symmetry grounds either; it depends upon the detailed model. A rough rule of thumb, based on experience in many different special cases, is that m-fold instability is more common for the smaller values of m. In particular a model of a rotating disc of self-gravitating matter (or of a circular drop of fluid held together by surface tension, which is very similar mathematically) has been studied by Debbie Lewis, Jerry Marsden, and Tudor Ratiu. They find that the initial instability is to two-fold symmetry. On the other hand, our opening example, the milk splash, is an analogous system with 24-fold rotational symmetry; so this argument is by no means conclusive.

Kinky Current

Spiral galaxies are perhaps the most spectacular instances of symmetry-breaking in astronomy – with one possible exception, which we postpone for a while because it is – literally – universal in scope. Among the other examples, one of the closest to home is a strange structure near Saturn's pole, known as Godfrey's kinky current. Saturn is a gas giant, a huge planet composed mainly of hydrogen and methane gases, perhaps with a small rocky core. Its atmosphere, like that of its larger cousin Jupiter, circulates around its axis in striped bands. The bands of Saturn are not terribly visible to the naked eye, but become obvious under computer enhancement. Most of what we know about them is due, again, to the *Voyager* spacecraft. Seen from the side, the bands appear to be horizontal straight strips. Seen from the poles, they are circular. Well, almost all of them. When D. A. Godfrey persuaded the computer to combine a number of *Voyager* photographs, and present the view looking down on the North Pole, he got a surprise. The band nearest the pole, instead of being circular, is hexagonal! The hexagon (Figure 6.8) rotates slowly, if at all, relative to the planet.

It's hard to doubt that whatever the precise physical mechanism behind Godfrey's kinky current, the mathematical mechanism is again symmetry-breaking. This time, the circular symmetry has become a six-fold symmetry. The phenomenon may be at least distantly related to the Earth's jet stream, which exhibits huge waves, often three- or four-lobed when viewed from the North Pole. But the

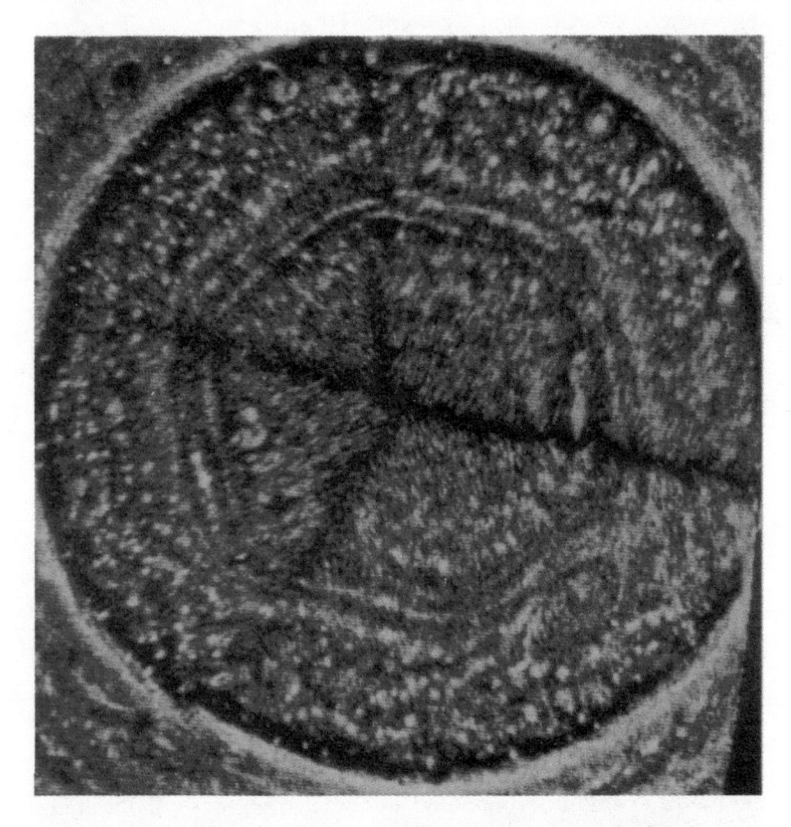

Figure 6.8 Godfrey's kinky current: a slowly revolving hexagonal structure at Saturn's north pole. The dark radial lines are artefacts of the way the picture was produced

jet stream's waves are time-dependent, swirly things, whereas Godfrey's kinky current is virtually in a steady state. Clearly there's a lot to be worked out here.

On a larger scale, pulsating stars provide more broken symmetries. The variable output of a cepheid, already mentioned in connection with intergalactic distances, is time-periodic, and thus an example of the breaking of time-translational symmetry, just as we have mentioned in chapter 3 in connection with Hopf bifurcation – the onset of a wobble. Indeed, one wonders whether some kind of Hopf bifurcation in stellar dynamics might be responsible. Some stars pulsate in ways that break their spatial spherical (or circular) symmetry. Our own Sun does this to some extent (Figure 6.9), although the wobbles are so small that they cannot be seen directly, and have to be inferred from spectroscopic measurements. These pulsations may well be examples of Hopf bifurcation with spherical symmetry.

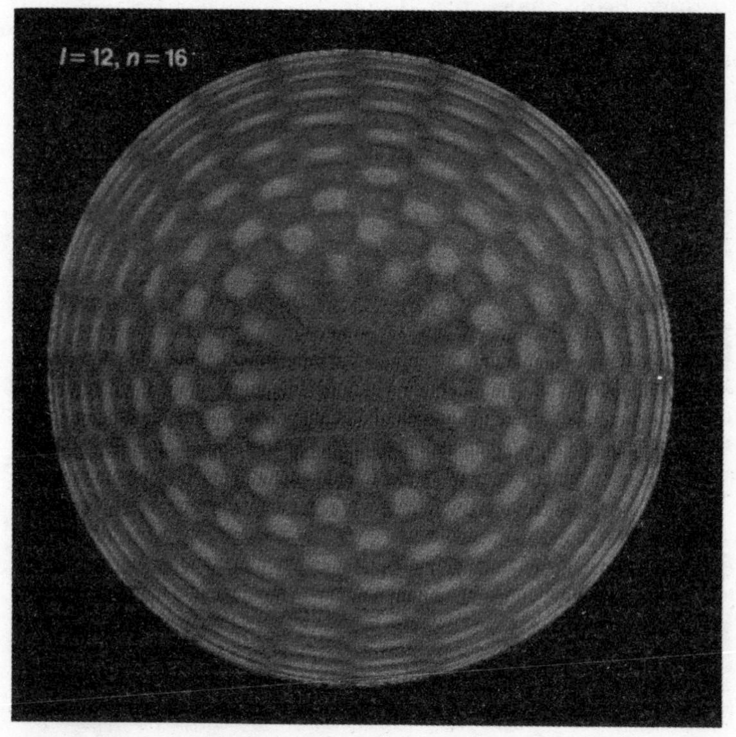

Figure 6.9 Symmetric oscillations of the Sun

The Great Wall

The most spectacular broken symmetry of them all is our own Universe. In a universe that is symmetric on large scales, there should always be the same density of galaxies in any region of space. However, computer analysis of surveys of galaxies carried out in the mid 1980s by Margaret Geller and John Huchra shows that they tend to collect together in vast filaments and sheets (Figure 6.10). Subsequent work has confirmed their findings, and led to the discovery of a number of huge voids. Geller and Huchra have recently announced the discovery of the largest coherent structure ever observed, a 'Great Wall' of galaxies 500 million light years long. Instead of being like a uniform 'galactic fluid', the universe is more like a mass of soap bubbles, with galaxies clustered on the surfaces of the bubbles, avoiding the holes inside. The reasons for this lack of uniformity are far from being understood. Computer simulations (Figure 6.11) suggest that gravitational instabilities are again responsible, and that they break the symmetry of a uniform distribution of galaxies. The idea is that as the universe expanded from the Big Bang, an initially symmetric distribution of matter was disturbed by random fluctua-

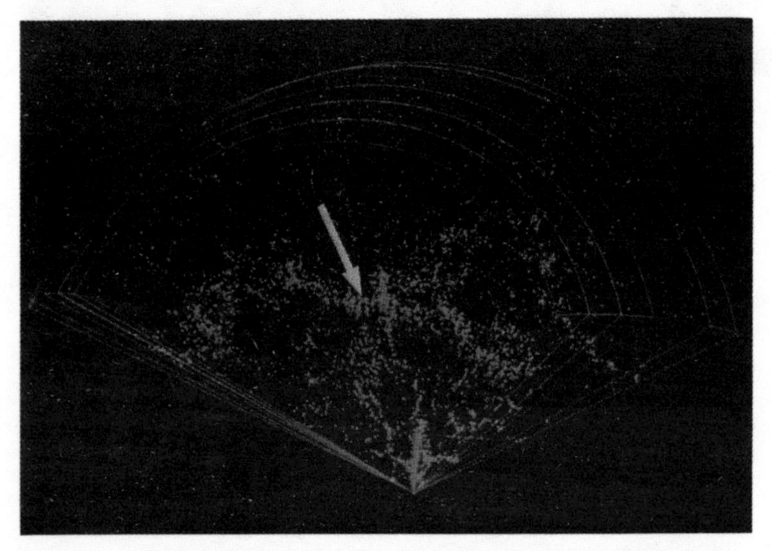

Figure 6.10 The Great Wall: a vast cluster of galaxies

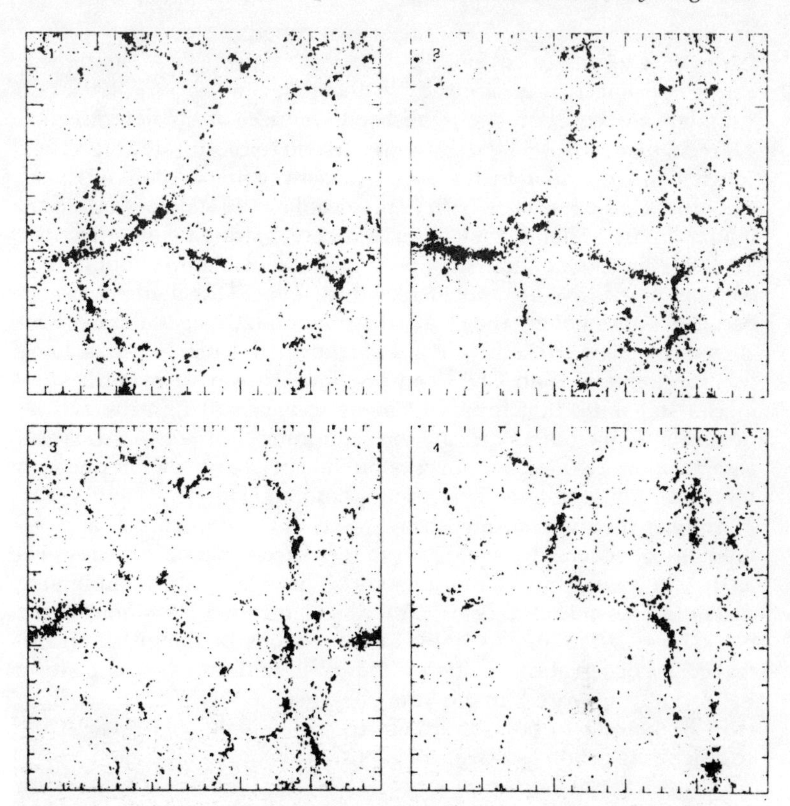

Figure 6.11 Computer simulation of galactic clustering, in which an initially random collection of galaxies clumps together under gravitational forces

tions, possibly quantum-mechanical in origin, and that gravity amplified these disturbances to create the filaments, sheets, and voids.

The snag in this theory is that the observed density of matter in the universe appears to be too small to produce strong enough instabilities. There simply hasn't been enough time for the voids to form by that mechanism. Here 'cold, dark matter' – matter made from unorthodox particles, matter that conveniently can't be seen through telescopes because it doesn't emit enough light – comes to the rescue, just as it did for the rotational data for galaxies. The CDM theory asserts that the extra mass of cold dark matter strengthens the gravitational instabilities. But the size of coherent structures like the Great Wall seems to be too big even for a universe containing this extra, invisible matter. Indeed, in the first few days of 1991 Will Saunders and collaborators published the results of a major survey of galaxies, using IRAS (the Infra-Red Astronomical Satellite). Their conclusion: 'the root mean square density variation falls off with smoothing scale less rapidly than predicted by the standard cold dark matter theory'. That is, they observe coherent structure on a scale too large for an explanation by CDM to be plausible. David Lindley, commenting in *Nature*, put it like this: 'The CDM model slips up because it does not get the galaxy distribution right on scales of about 20 megaparsecs [80 lightyears] and above: the real sky has more non-randomness than CDM can provide.' However, many cosmologists still think that the CDM model can be put right by further tinkering, while others dispute the significance of the IRAS data and assert that more extensive surveys provide better agreement between theory and observation. This one will run and run.

Another theory is 'cosmic string'. A cosmic string is a defect in the topology of spacetime, analogous to the defects that occur in crystals as they grow, when different regions of the crystal lattice don't quite match up. Cosmic strings are rather like elongated black holes. Like black holes, they are currently the inventions of theorists, unconfirmed by observation. Evidence that *might* indicate a cosmic string has recently appeared, in the form of a cluster of 'twin' galaxies. Each twin is thought to be a double image of a single galaxy, with the cosmic string action as a distorting lens. Who knows? It could be true. Anyway, a very long cosmic string is unstable, and tends to break up into smaller loops (itself a symmetry-breaking instability not unlike the formation of drops of dew on a spiderweb). These loops might trigger the formation of voids and sheets of galaxies.

Whatever the explanation turns out to be, there is broken symmetry in the universe, on the grandest of scales. It strikes at the heart of current cosmological thinking.

7

Turing's Tiger

Tyger! Tyger! burning bright
In the forests of the night,
What immortal hand or eye
Could frame thy fearful symmetry?

William Blake, *The Tyger*

Symmetries abound in the world of living creatures. We ourselves are, to a good degree of approximation, bilaterally symmetric: a human being seen in a mirror still looks like pretty much the same human being, which is why we get in such a tangle when we ask why mirrors reverse right-left but not up-down. Depending on your point of view, they do both – or neither. If we were bilaterally symmetric we wouldn't think of our image in a mirror as another person, and then we wouldn't think that our image's left hand is the other person's right. (Indeed, on the planet Erlgray there are two alien races, the T'pots and the Top'ts, the first of which resembles a teapot with eyes to the front, a spout on its left, and a handle on its right, while the other race is its precise mirror image. The two are completely separate species; they can't interbreed, although for reasons of delicacy we'll leave the explanation to your imagination. In fact, they're deadly enemies. When a T'pot looks at itself in a mirror (Figure 7.1), it doesn't wonder why its left and right appendages have been interchanged: it wonders why it has been transformed into a ferocious Top't.)

William Blake's tiger is something of a cliché in books about symmetry, but it's impossible to avoid the creature. According to Martin Gardner, the geometer H. S. M. ('Donald') Coxeter has suggested that Blake was referring to its bilateral symmetry. We beg to differ, as our unfolding tale (tail?) will elucidate. Biological symmetry

Figure 7.1 T'pots and Top'ts. Why does a mirror change allies into enemies?

isn't perfect: for example the left-hand side of your face is not precisely the same as the reflection of the right-hand side; more systematically your heart is on the left side of your body. But these all occur within an overall symmetric form. Starfish, unsung by Blake, are also symmetric, but their symmetry is that of a regular pentagon. An octopus has a rather superficial eightfold symmetry. Flowers have symmetric arrangements of petals, and surprisingly often the number of petals belongs to the so-called *Fibonacci sequence*

1, 2, 3, 5, 8, 13, 21, 34, 55, 89,...

in which each number is the sum of the previous two. Viruses often have the symmetry of the icosahedron (Figure 7.2).

Behaviour, as well as form, also involves symmetric structures. Bees build honeycombs arranged like a hexagonal lattice, and similar structures occur in wasps' nests. In chapter 4 we saw that the areas of lake-bed controlled by territorial fish can also form hexagonal lattices. Yet, in the early stages of development, living creatures are just tiny egg cells, often roughly spherical. And when a hive of bees sets out to build a honeycomb, it starts with a more or less flat surface. Moreover, we think of life as an organic, flexible process, whereas symmetry is a kind of structural rigidity. Where, then, do these symmetries associated with living creatures come from?

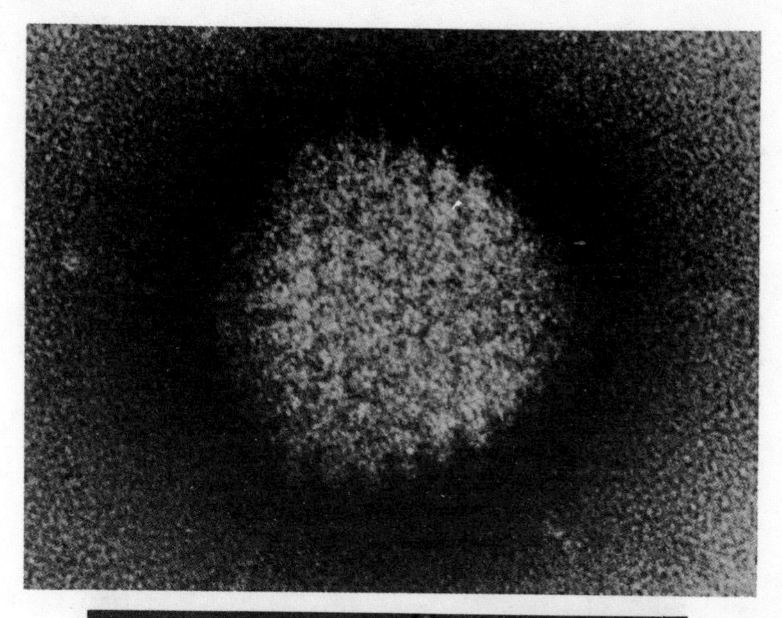

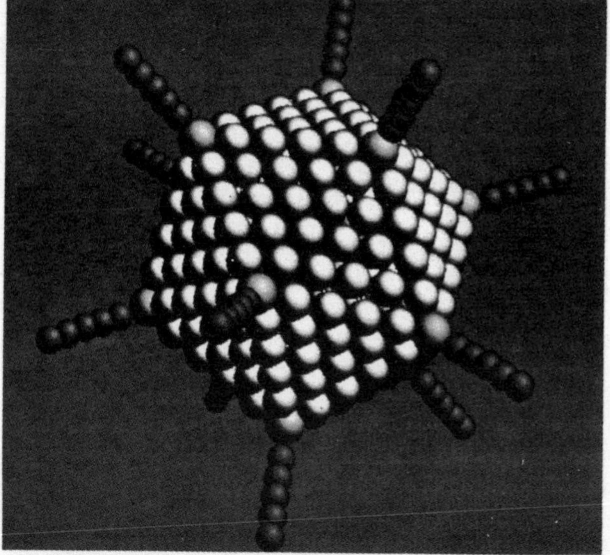

Figure 7.2 Icosahedral structure of the Adenovirus

Consider a Spherical Frog

The only symmetry in the form of an adult frog is bilateral. This bilateral symmetry was firmly established long before the frog became a tadpole. But earlier in its development – in fact at two distinct stages – the frog was spherically symmetric, and at two related stages it was circularly symmetric (Figure 7.3). Imagine it, a circular frog! Of course a great deal more goes into frog development than just a series of changes of symmetry, and later we'll peep into Pandora's developmental biological box; but for the moment let's follow our noses and concentrate on observable symmetries of form.

Frogs, like us, begin their existence as a single fertilized egg cell, or *zygote*. The cell is approximately spherical in form, so the frog embryo starts its life possessing external spherical symmetry. However, the spherical form is to some extent misleading, because the yolk is distributed unevenly, being concentrated towards one end. The cell then divides repeatedly, a process known as *cleavage*, and the first few stages of this break the spherical symmetry piece by piece until nothing is left. The first stage of cleavage, into two cells, creates a symmetric pair of rounded hemispheres, separated by an almost flat membrane. An axis of circular symmetry runs down the middle, and the pair of cells is symmetric under all rotations about this axis, and all reflections in planes containing this axis. In addition, it's symmetric under a reflection in the plane at right angles to the axis, the plane that divides the two cells.

The two cells cleave again, becoming four. Cells are rather like blobs of sticky jelly, and when they divide they don't always stay in place – in many animals they slide around and rearrange themselves. However, frog cells don't do this, so the symmetry of the resulting structure is essentially that of the square, although in practice two of the cells are a bit larger than the other two.

In mammals, where cells can slide around, the symmetry is different: the same as a tetrahedron. Think of four ping-pong balls placed at the corners of a tetrahedron. One way to see them is as two pairs of balls, with one pair sitting at right angles to the other. Now imagine inflating the balls so that they come into contact with each other: that's what the four mammalian cells look like.

In either case the next division, into eight cells, breaks most of the symmetry of the four-cell stage. Cleavage continues, and soon there's no symmetry left at all. The precise sequence of steps varies from creature to creature (Figure 7.4). In some, especially nematode worms, the process is quite weird. Cell-divisions aren't synchro-

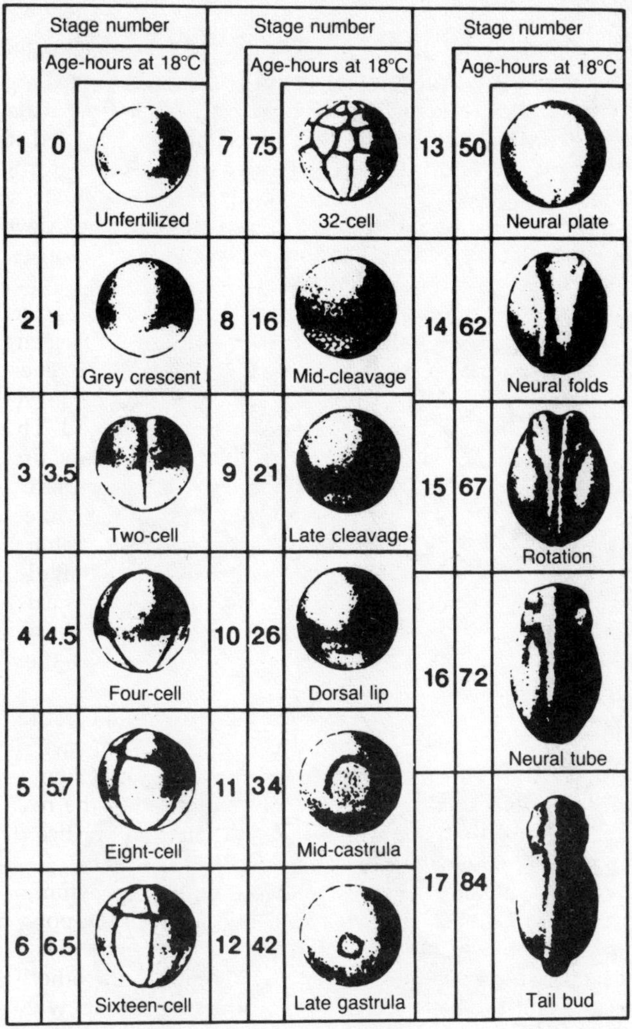

Stage number	Age-hours at 18°C		Stage number	Age-hours at 18°C		Stage number	Age-hours at 18°C	
1	0	Unfertilized	7	7.5	32-cell	13	50	Neural plate
2	1	Grey crescent	8	16	Mid-cleavage	14	62	Neural folds
3	3.5	Two-cell	9	21	Late cleavage	15	67	Rotation
4	4.5	Four-cell	10	26	Dorsal lip	16	72	Neural tube
5	5.7	Eight-cell	11	34	Mid-castrula	17	84	Tail bud
6	6.5	Sixteen-cell	12	42	Late gastrula			

Figure 7.3 Frog development: note changes in symmetry

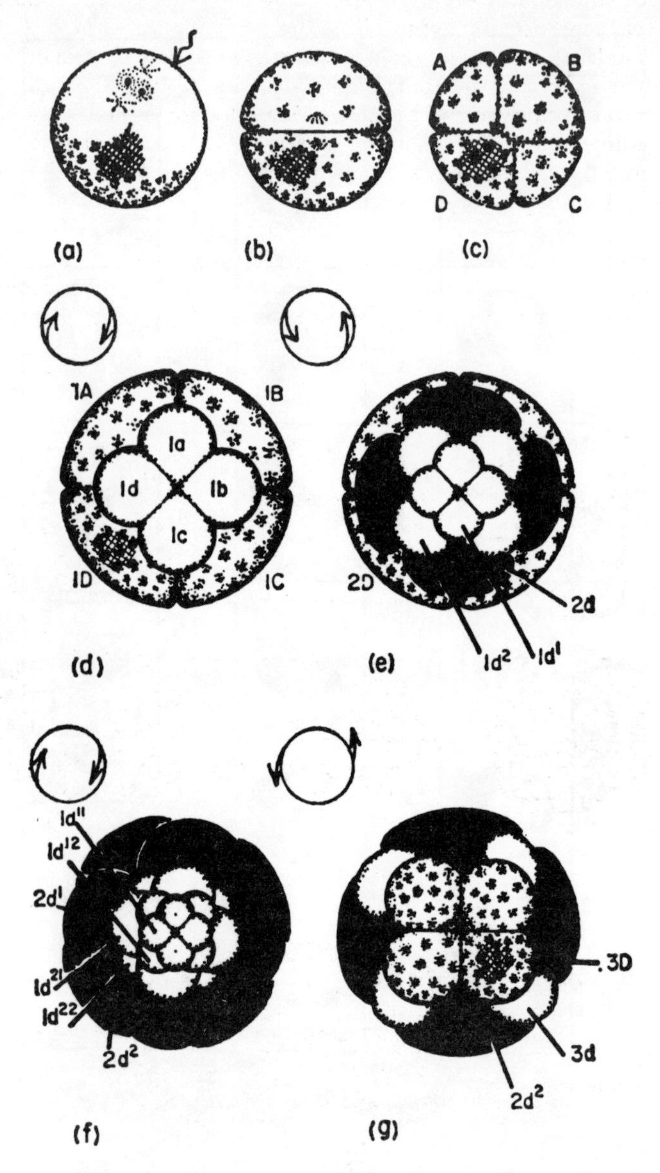

Figure 7.4 Early stages in the sequence of cell divisions, shown here for polychaete worms

nized, so the numbers don't keep doubling: there is for example a six-cell stage, (c) in Figure 7.5. The same figure shows that for nematodes symmetry tends to be lost sooner than it is for many other animals. The cells in adult nematodes can be matched with each other, one for one: it looks as if all nematodes are kit-built, assembled according to the same rigid genetic rules. This is fascinating, but it suggests that nematodes are exceptional creatures, and we'll concentrate on animals whose development involves a certain amount of dynamical freedom. We should add that for some animals the cells tend to cleave into equal-sized pieces, while others have large variations in the size. These size differences can affect the symmetries that we observe. So for the moment let's confine our attention to frogs.

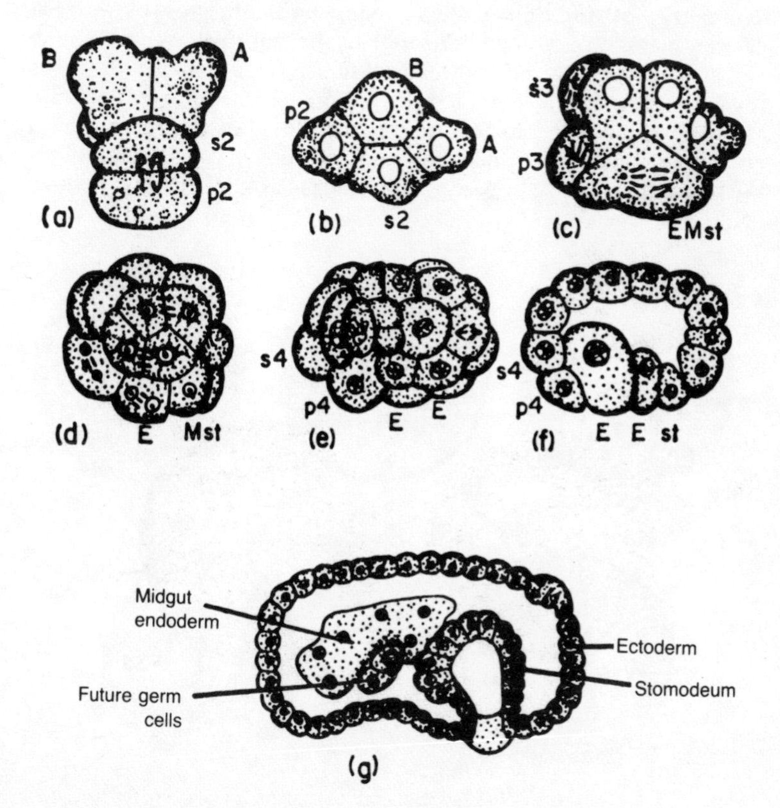

Figure 7.5 Cell divisions in nematode worms

As the cells increase in number, they become smaller. Soon they form a hollow shell, called the *blastula*, stages 8–9 in Figure 7.3. Strictly speaking, the blastula has no symmetry, if we take very fine detail of the shapes of component cells into account; but the overall shape of the blastula is that of a hollow sphere. So, to a good approximation, the embryo has regained spherical symmetry.

The next few stages are fundamental to the form of the final adult frog. The first of them is called *gastrulation* ('stomach-formation') and is shown in stages 10–12 of Figure 7.3. The hollow sphere of the blastula begins to collapse inwards upon itself. Mathematically, the situation is analogous to a spherical shell buckling under pressure. We've already seen what happens in this mechanical system: the buckled shell is circularly symmetric, with a depression at one end of its axis of symmetry. An analogous phenomenon, with similar symmetry but differing in detail, happens in the frog blastula. A circular mark forms somewhere on the spherical shell, rather as if the egg is being cut by an invisible knife, and the blastula collapses inwards (Figure 7.6). It starts to look more like a short fat tube; and already we can begin to see the form of the adult emerging, for all animals are highly decorated and modified versions of short fat tubes. We might imagine that the reason for the collapse is that the 'hollow'

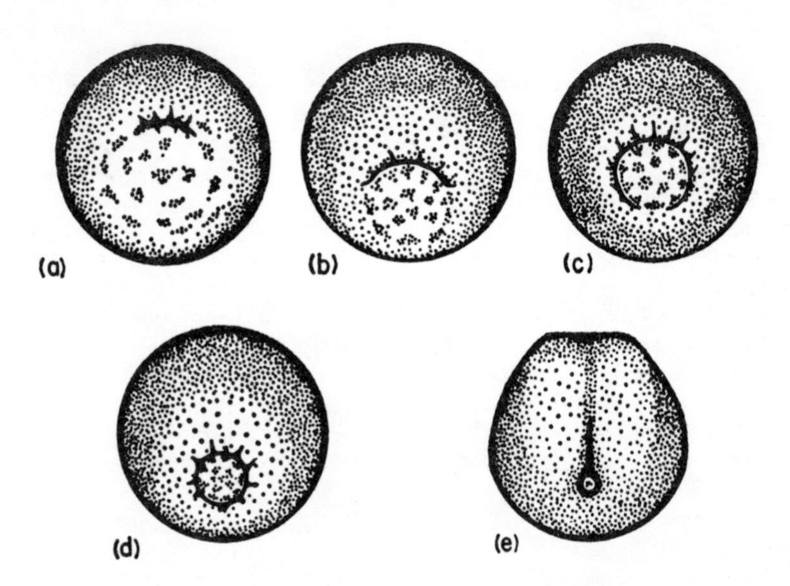

Figure 7.6 Gastrulation in the frog

blastula isn't empty: it's like a hollow torus, filled with fluid containing all kinds of chemicals. It floats in a 'sea' of fluid, also containing chemicals. Could this set-up create a kind of osmotic pressure-difference between the inside and outside of the blastula? If so, gastrulation is basically a mechanical phenomenon! However, it turns out that nature is rather more subtle: we explain just what does happen later in this chapter.

The second crucial stage of development is *neurulation* ('nerve-formation'), the first appearance of what will eventually become the backbone with its enclosed spinal cord, the main trunk line of the nervous system. See stages 14–16 of Figure 7.3. In the earliest stages of neurulation, the circularly symmetric tubular structure of the embryo distorts and flattens, becoming bilaterally symmetric. This bilateral symmetry is never broken again, at least not as regards the gross form of the animal: when limbs develop, they develop similarly on both left and right; when eyes develop, there is one on each side, and so on. Some internal organs, such as the kidneys, form symmetrically. However, the development of some of the frog's internal organs, such as the gut, breaks even the bilateral symmetry, just as it does with us.

The changing symmetries of the frog embryo set a pattern that, in broad outline, but with modifications at some stages, is common to many animals – fish, amphibians, mammals, even insects. Other creatures also follow patterns of changing symmetry, but with different results. A developing starfish, instead of becoming bilaterally symmetric, ends up with approximate five-fold symmetry. As we know, that's another of the possible ways in which circular symmetry can break. There's no mathematical reason why creatures with six-fold symmetry, or seven-fold symmetry – or indeed 73-fold symmetry – shouldn't similarly develop. Many such symmetries can be seen in the petals of flowers, the tentacles of sea-creatures, the fronds and fringes of microscopic pond life. Even so, some patterns are far commoner than others: nature seems to have some favourite patterns, and not everything that's mathematically possible necessarily happens.

Designer Genes

How do these patterns of development arise, and why do we find typical mathematical symmetry-breaking sequences among them? We live in an age that has discovered a remarkable analogy between genetics and computation. The developmental path of an embryo is controlled, to a great extent, by its genes. The genes contain a chemical molecule known as DNA, which forms long helical strands. The strands are tied together by four different molecules, known as *nucleotides*: adenine, cytosine, guanine, and thymine. These four nucleotides are like symbols in a language, and the sequences that they form contain the instructions for making the protein molecules whose properties prescribe the form of the developing creature. So DNA can be seen as a kind of computer program, which specifies the structure of the developing embryo. That poses a problem: from the 'program' point of view it's hard to see *why* symmetries should occur at all in living creatures. We'd be tempted to look for symmetries within the DNA program, or at least for instructions that create symmetries. Such a search would almost certainly prove fruitless, for there is little reason why a program should follow the natural mathematical pattern for changes of symmetry in dynamical systems. A program can contain a pretty arbitrary set of instructions, it doesn't have to mimic dynamics. It can build the organism cell by cell, like someone knitting socks from a pattern.

However, biological development is not simply a matter of 'coded' instructions. Those instructions have to be acted upon. Indeed they only possess meaning within a context that can put them into action, just as a computer program is useless without a computer to run it. How is the DNA program run? This is the 'biggest unsolved problem in biology', and nobody really knows the answer. However, they've discerned some of the basic outlines.

One thing that's clear is that the DNA 'instructions' don't specify the process in complete detail. They're not like a knitting-pattern, which tells you where every stitch must go. The instructions in DNA don't say things like 'switch cell number 100325 to muscle-making mode and send it to location 99432 in the left arm'. We can be sure of this because the DNA molecules in living creatures aren't long enough, they don't contain enough 'information', to do things that way. If they did, the instruction manual for assembling a living creature cell by cell would have to use much more DNA than actually occurs. If somebody tried to sell you a knitting-pattern for a complicated sweater, printed on a postage stamp in normal sized letters, you'd be suspicious.

Moreover, development is not just a matter of sending a DNA message from transmitter (parent) to receiver (offspring). For a start, the supposed 'receiver' doesn't exist when the message is 'sent'. A more accurate statement would be that the message is sent from the parent to itself, and it tells the parent organism how to develop the offspring. A great deal of the process relies upon what's already present in the parent, as well as on 'instructions' coded in the DNA. Mammal DNA is shorter than amphibian DNA, it contains fewer instructions! But that doesn't mean that a frog 'contains' or 'requires' more information than a chimpanzee. Much of the frog's DNA is there to specify alternative development paths for different environmental factors, such as temperature of the pond in which the frog is growing. A developing chimpanzee, however, is kept at a constant temperature, because it is still inside its mother. By putting instructions for maternal temperature-regulation into the chimpanzee genetics, nature has managed to eliminate a much more extensive set of instructions for dealing with a changing environment. As an extreme example, the DNA 'information' in some species of amoeba is about thirty times as large as that in humans!

The developmental process, then, is not specified completely by the coded data in the DNA. It relies upon a context; in particular, the parents contribute far more than just their DNA sequences. But there's a greater context, which is also important, and it's here that the mathematics of dynamical systems, and the phenomenon of symmetry-breaking, find a natural role. That greater context is the laws of physics and chemistry. The DNA program does not have to contain instructions such as 'make salt crystals cubical' or 'make this type of fat molecule hydrophobic'. These properties follow inevitably from the laws of physics, and you can't avoid them even if you want to.

Epigenesis

The picture of development that emerges from all this is called the theory of *epigenesis*, and was introduced by Caspar Wolff in the second half of the eighteenth century. The idea is that the fertilized egg contains a set of instructions that are sufficient to determine how the adult should be built. (The 'knitting-pattern theory', in contrast, would require it to describe where every stitch goes.) The developing embryo is a complicated, spatially distributed, time-varying dynamical system, which behaves in an intricate manner according to the laws of physics and chemistry. The instructions in DNA set the system moving in some particular manner – push it in a desired direction, so to speak – and then wait for it to develop, following the physical laws. (Development is *not* a simple egg-begets-chicken process: the cell nuclei are 'passengers' in the early part of development, and initially the 'instructions' come from the *parent*, but at a later stage the DNA in the zygote takes charge. The break between the generations does not occur at fertilization, but later.) Every so often the system needs another push to keep it on the right track: then the genes come into play again. Thus what is coded in the DNA is not a complete *description* of where every cell must go and what it must do, but a *prescription* of what must be done to the dynamical system to control its development in the appropriate manner.

Think of telling someone how to fill a pond with water. It isn't necessary to list precisely where each water molecule must be sent. It's sufficient to instruct them to point a hose in the right direction and turn on the tap. The laws of physics take care of everything else. In a similar manner, you can direct a stream of water down a complicated hillside into a particular valley just by letting it flow, but nudging it gently from time to time near places where its path might branch away from the desired one. This kind of image is due to C. H. Waddington, who called it the 'epigenetic landscape'.

The place where the mathematics of dynamical systems is given free rein is in the gaps of dynamic evolution that occur between the DNA-triggered switches and control interventions. So if an initially spherical system (single cell or blastula) is left to run on its own dynamics, it will either stay spherically symmetric, or break symmetry in the expected mathematical manner. It may not be necessary to code in the DNA for every detail of the collapse of the blastula: some aspects may be the result of simple dynamical forces. But it is necessary for the DNA to be perpetually meddling with the dynamics, stabilizing this state, selecting that one, compensating for an environmental change, making some cells grow into kidneys and others into bone, and so on.

In fact, even this is too simplified a picture; for there's no reason why the two processes of DNA 'switching' and dynamical flow can't go on *simultaneously*. They will affect the development in very different ways, to human eyes, and it's therefore convenient to *think* of them as separate processes; but that distinction need not be an either/or one. A good analogy is the behaviour of seawater, which is governed both by the dynamics of waves and by the chemistry of its constituents. They work in different ways, but real seawater combines their influences into a single result. For instance, you must have noticed how spindrift – frothy masses of foam – can form in turbulent seawater. What causes it? It looks as if the dynamics of turbulent flow should be responsible – we know turbulence causes frothing, just turn on a tap. But why does the froth *persist*? Froth in tapwater dies away very fast – unless you add washing-up liquid. Spindrift is composed of bubbles caused by turbulence, but those bubbles are stabilized by proteins of biological origin. The protein molecules

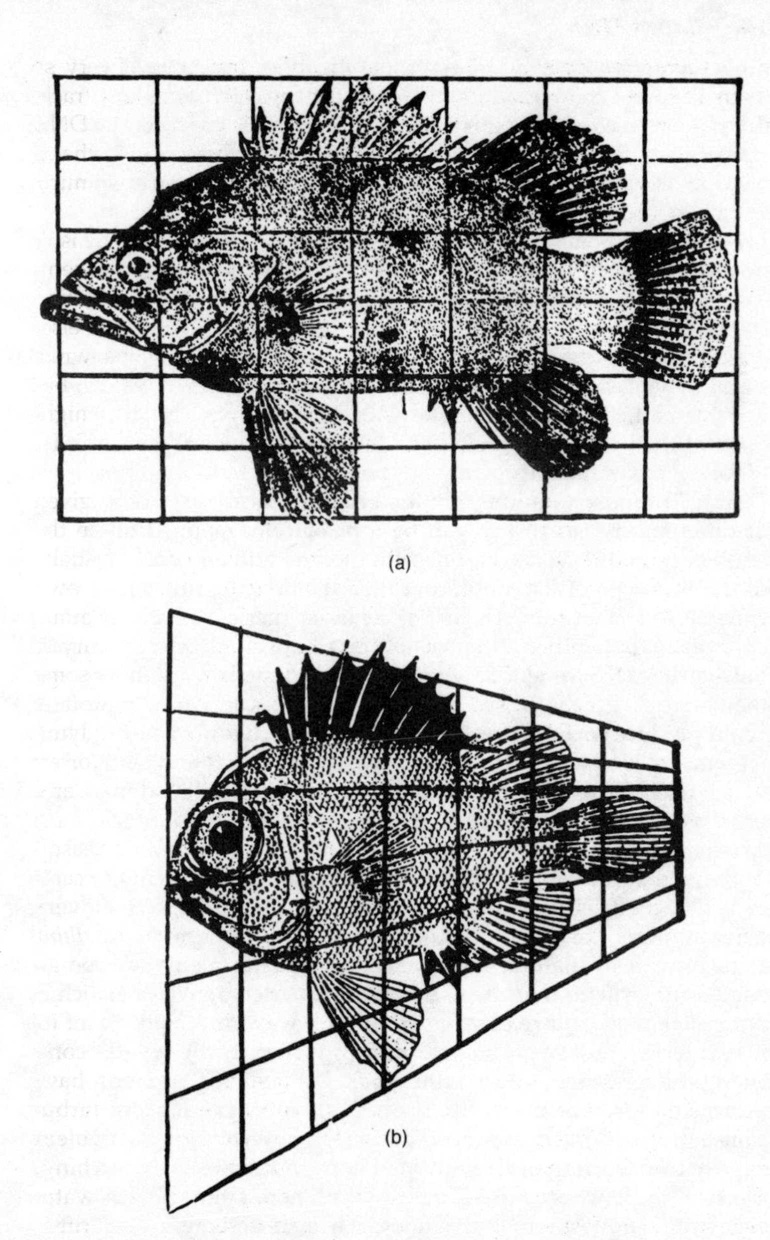

Figure 7.7 Topological transformation mapping the outward form of one species of fish, Polyprion *(a) to that of another,* Pseudopriacanthus altus *(b)*

attract water at one end and repel it at the other – just like detergent. They separate the bubbles and prevent them combining into larger bubbles that would be unstable, and burst. So the two 'separate' processes of fluid dynamics and chemistry unite to create spindrift. If astronauts were to land on a planet of a far star and observe spindrift in the oceans, they would know immediately that life must be present. The chemistry of seawater is analogous to the role of DNA in development; the fluid dynamics of seawater is analogous to the 'free-running' dynamics of the cell. *Genes aren't for shapes, they're for chemistry*.

Within the still greater context of evolution, we can see biological development as an opportunistic process. It takes advantage of certain mathematical possibilities – or at least, physical manifestations of mathematical possibilities – for the dynamical development of form, makes them its own, and moulds them to its own purposes. D'Arcy Thompson noticed that the gross physical forms of a variety of different species of fish can be topologically mapped on to one another (Figure 7.7). If you start with the first grid and draw a picture of the first type of fish on it, and then distort the grid as indicated, then the form that results is that of the other species. Relatively minor changes in dynamics could plausibly be responsible for such similarities – just let the system flow for a bit more time here, a bit less time there – but it's hard to see how small changes in DNA 'programs' could produce such an effect. In this book, which is not a biology text, we emphasize general features of the dynamical foundations upon which true biological development may be built. But the dynamics is only a tiny part of a much more complicated and much more wonderful story.

There may be evolutionary advantages in having symmetry. In fact, if Darwin is right, there must be. By definition, a symmetric object repeats the same structure many times. So symmetric dynamics provides a natural mechanism for making many copies of a successful structure, such as a limb or a vertebra. Whether or not nature takes advantage of symmetry in this way is not fully clear; but it is true that for a very long period of evolution the world's most successful creatures were arthropods. Arthropods are symmetry incarnate: segmented creatures which, like centipedes, repeat the same structure over and over again along their length. You can specify the structure of an arthropod very economically, which saves a lot of work when coding their DNA: 'here's how to make one segment ... now repeat forty times'. Of course some extra instructions are also needed for the head and tail, which don't follow the symmetry rule: here the DNA code must intervene to prevent the natural dynamics from producing an infinitely long animal. As we said, evolution is opportunistic.

Morphogens and Morphogenesis

Let's pursue this idea, that a developing embryo obeys 'natural dynamics', at least when its genetic programs aren't tinkering with its biochemistry. So far what we've done has been descriptive: we've looked at how development occurs and noticed that it involves changes of symmetry. Can we give this idea predictive value, so that it tells us where to look for new phenomena?

D'Arcy Thompson studied the symmetries of developing creatures in *On Growth and Form* in 1917, pointing out that 'symmetry is highly characteristic of organic forms, and is rarely absent in living things'. He introduced a dynamical interpretation of symmetry, and traced its origin in living creatures to the fact that equilibrium configurations often possess symmetry. In support he quoted the physicist Ernst Mach:

> In every symmetrical system every deformation that tends to destroy the symmetry is complemented by an equal and opposite deformation that tends to restore it. ... One condition, therefore, though not an absolutely sufficient one, that a maximum or minimum of work corresponds to the form of equilibrium, is thus applied by symmetry.

This rather obscure statement is probably a forerunner of the 'principle of symmetric criticality', which states that when seeking a symmetric equilibrium it is unnecessary to consider perturbations that break the symmetry. Olga Ladyzhenskaya has set this principle on firm mathematical foundations.

A more detailed theory of developmental symmetry was published in 1952 by Alan Turing. We shall spend some time talking about it – not because it is *right*, because in detail it almost surely isn't, but because it is a good way to approach some very important general ideas. Turing – perhaps best described as a mathematical logician but, as we shall see, far more versatile than that – is mainly remembered for his contributions to both the theory and the practice of the electronic computer. In 1936 he introduced what is now known as a 'Turing Machine' – a mathematical idealization of the computational process that involves reading digits 0 or 1 from an infinitely long tape and writing them back according to predefined rules. His 'Turing test' for intelligent behaviour is much favoured by the Artificial Intelligentsia (because it keeps open the hope of creating intelligent computers) and attacked by their opponents (for the same reason). During the second world war he was based at the Code and Cypher school in Bletchley, England, and his job was to break the German High Command's code messages. The sinking of the *Scharnhorst* on 26 December 1942 was due in part to his group's efforts. Writing codes on a regular basis is very time-consuming, so the Germans had developed machines (the ENIGMA being a famous one) to perform the task. Breaking codes is even more time-consuming, and Turing's group became more and more convinced that what they needed was an electronic data-processing device – in short, a computer. In 1943 they designed and built one, known as Colossus: the whole process took only eleven months! It contained 1,500 vacuum tubes, and it included many features characteristic of today's computers, such as a clock pulse to keep all operations in step with each other, binary circuits, and – at Turing's insistence – conditional branching of programs. Colossus was arguably the first electronic programmable computer, even though that honour is traditionally accorded to ENIAC, built in 1946 under the direction of Presper Eckert and John Mauchly at the University of Pennsylvania. One reason for the historical neglect of Colossus may be that its details were revealed only in 1975 because of the British government's 30-year rule on the release of sensitive documents.

Turing was a brilliant mathematician and logician, with a somewhat awkward personality: something of a loner. He died of poisoning, and the coroner recorded a verdict of suicide. Douglas Hofstadter describes him as 'highly unconventional, even gauche in some ways, but so honest and decent that he was vulnerable to the world', and that about sums it up.

Turing's 1952 paper displays his versatility, for it is in a field far removed from computation and mathematical logic. Its title is *The Chemical Basis of Morphogenesis*. Morphogenesis ('shape-generation') is the process whereby a developing embryo acquires form. Turing argued that:

> a system of chemical substances, called morphogens, reacting together and diffusing through a tissue, is adequate to account for the main phenomena of morphogenesis.

In particular, he discussed the development of patterns:

> Such a system, although it may originally be quite homogeneous, may later develop a pattern or structure due to an instability of the homogeneous equilibrium, which is triggered by random disturbances.

Sounds familiar?

Turing acknowledged that his theory was 'a simplification and an idealization, and consequently a falsification'. He never expected it to to account for all features of morphogenesis. But he felt that it captured some of the important behaviour. We'll see later that there are definite difficulties with Turing's ideas if they are taken too literally: biology is subtle and devious, and astonishingly complex. But his theory does offer valid and important insights into the development of biological form: in particular the idea of pattern-formation through symmetry-breaking.

For Turing's purposes, the cellular structure of the developing organism is ignored: instead, it is considered as a more-or-less homogeneous mass of tissue. A biologist will justifiably protest that this throws out all of the interesting biology! No matter: it can be added back later. Turing's real question is 'to what extent are the possibilities for morphogenesis independent of biological detail?' To answer that, we *have* to remove the biology.

Chemical substances are diffusing through the tissue, and reacting together: these are Turing's 'morphogens'. The distribution of chemicals within the tissue changes with time, according to dynamic laws. We might imagine the tissue wandering through a 'chemical landscape', propelled by the forces of reaction and diffusion: thus Turing's model can be viewed as a simplified chemical version of Waddington's epigenetic landscape. Each position in the landscape corresponds to a particular distribution of chemicals (and concentrations thereof) within the tissue. The forces of reaction and diffusion provide a dynamic that determines the tissue's motion around the landscape.

Turing's view is by no means as naive as it might seem to be. In particular, he insists that morphogens do not directly produce the form of the organism. Instead, they trigger changes in that form. He refers to 'evocators', a term introduced by Waddington in 1940: 'These evocators diffusing into a tissue somehow persuade it to develop along different lines from those which would have been followed in its absence.' Other examples that he cites are hormones and perhaps skin pigments. He considers but rejects the idea that the genes themselves may be morphogens, on the grounds that they do not diffuse. Later we'll return to these ideas and discuss their relation to current biological theories; but for now we'll accept Turing's simplifications, and see where they lead.

The Breakdown of Symmetry

Turing recognizes, quite explicitly, the role of symmetry-breaking. We quote the relevant section of his paper at length because it is so apposite. He was decades ahead of his time.

> 4 THE BREAKDOWN OF SYMMETRY AND HOMOGENEITY
> There appears superficially to be a difficulty confronting this theory of morphogenesis, or, indeed, almost any other theory of it. An embryo in its spherical blastula stage has spherical symmetry, or if there are any deviations from perfect symmetry, they cannot be regarded as of any particular importance, for the deviations vary greatly from embryo to embryo within a species, though the organisms developed from them are barely distinguishable. One may take it therefore that there is perfect spherical symmetry. But a system which has spherical symmetry, and whose state is changing because of chemical reactions and diffusion, will remain spherically symmetric forever. (The same would hold true if the state were changing according to the laws of electricity and magnetism, or of quantum mechanics.) It certainly cannot result in an organism such as a horse, which is not spherically symmetrical.
>
> There is a fallacy in this argument. It was assumed that deviations from spherical symmetry in the blastula could be ignored because it makes no particular difference what form of asymmetry there is. It is, however, important that there are some deviations, for the system may reach a state of instability in which these irregularities, or certain components of them, tend to grow. If this happens a new and stable equilibrium is usually reached, with the symmetry entirely gone. The variety of such new equilibria will normally not be so great as the variety of irregularities giving rise to them. In the case, for instance, of the gastrulating sphere, discussed at the end of this paper, the direction of the axis of the gastrula can vary, but nothing else.
>
> The situation is very similar to that which arises in connexion with electrical oscillators. It is usually easy to understand how an oscillator keeps going when once it has started, but on a first acquaintance it is not obvious how the oscillation begins. The explanation is that there are random disturbances always present in the circuit. Any disturbance whose frequency is the natural frequency of the oscillator will tend to set it going. The ultimate fate of the system will be a state of oscillation at its appropriate frequency, and with an amplitude (and a wave form) which are also determined by the circuit. The phase of the oscillation alone is determined by the disturbance.
>
> If chemical reactions and diffusion are the only forms of physical change which are taken into account the argument above can take a slightly different form. For if the system originally has no sort of geometrical symmetry but is a perfectly homogeneous and possibly irregularly shaped mass of tissue, it will continue indefinitely to be

homogeneous. In practice, however, the presence of irregularities, including statistical fluctuations in the numbers of molecules undergoing the various reactions, will, if the system has an appropriate kind of instability, result in this homogeneity disappearing.

Arguments like these are bread and butter to us by now, because we've come across them so many times already. But it's worth going through Turing's discussion in a little more detail, to show just how many important ideas are present in it. Remember, this was 1952, some forty years ago.

What's bothering him is an objection along the lines of Curie's Principle, in its narrow interpretation of symmetry-conservation – which of course we know is false. It seems likely that he included this discussion because biologists had raised precisely this objection, not realizing its falsity. Turing proceeds to argue that – phrased in more modern language – when the symmetric equilibrium state loses stability, random fluctuations can trigger a symmetry-breaking bifurcation. He gives two examples, both illuminating. The first is gastrulation, which we've just discussed and to which we'll return below. Turing's point is that the typical bifurcation from spherically symmetric states is to axially symmetric states: not all symmetry is lost, and the only arbitrary choice is the position of the axis. His second example is a physical manifestation of Hopf bifurcation (see chapter 3), the onset of time-periodic oscillations, or wobbles. We'll see in the next chapter that the analogy between electronic circuits and biological development is astonishingly fruitful; but the important point to observe here is that Turing recognizes Hopf bifurcation as a form of temporal symmetry-breaking.

Or does he? First, let's adopt a Devil's Advocate position. This is that Turing is really just talking about bifurcation, and he is not making its symmetry-breaking features explicit. His discussion is mostly about growing instabilities triggered by small random events. That's true, but it ignores one crucial sentence, the discussion of the phase of the oscillation as its only arbitrary feature. The phase is clearly portrayed as playing a role analogous to the symmetry-axis of a buckled sphere. Recall the implications of the Extended Curie Principle: when a sphere buckles to an axisymmetric state, the axis is arbitrary, but all possible buckled states are identical except for rotations of the axis. Similarly, when a steady state of an electronic circuit begins to oscillate, the phase is arbitrary, but all possible oscillatory states are identical except for translations of the phase. Solutions that break symmetry occur in symmetrically related bunches. The idea of phase-shift symmetry in periodic oscillations is implicit in Turing's discussion.

The final paragraph quoted from Turing's paper introduces yet another key idea: that symmetries may be local things, not necessarily related to the overall form of the object concerned. Homogeneity is a kind of symmetry, even if it occurs in an object that doesn't have a symmetric form. The kind of 'symmetry operation' we have in mind is this: cut two small balls out of the mass of tissue, and swap them. If the tissue is homogeneous, this operation makes no difference. If it's not homogeneous, then changing balls whose contents differ does make a difference, so homogeneity is equivalent to this 'local symmetry'. Turing is saying that even in an irregularly shaped mass of tissue, initial homogeneity is a type of symmetry, and it too can be broken.

Incidentally, the two balls that we cut out might be the same: that is, we cut out a ball and replace it. If we are allowed to rotate it before we replace it, then 'symmetry' implies that the state of the tissue is the same in any direction, that is, it is *isotropic*. So isotropy is local rotational symmetry.

Hydra Dynamics

Turing works out several examples in his paper. The fundamental example is a circular ring of cells, see figure 3.7. We noted there that this system has dihedral group symmetry, the same symmetry as a regular polygon whose vertices are the cells and whose edges join neighbouring cells. In chapter 3 we described some recent discoveries about such systems. Turing anticipated some of them, finding two main types of solution: time-independent stationary waves, and time-dependent travelling waves. In fact more recent work has shown that there can also be time-dependent standing waves, together with a whole host of more complicated solutions up to and including chaos! But Turing's solutions are among the commonest actually encountered.

What do these solutions look like? For the stationary wave solution, the morphogen concentrates in equally spaced clumps around the ring. The clumps become more pronounced with time, but they don't move. The symmetry that remains is that of a regular polygon, but generally with fewer vertices than the number of cells; in fact, a divisor of that number. So a ring of 12 cells, for instance, might have 1, 2, 3, 4, 6 or 12 clumps. The travelling waves are similar but now the entire system of clumps rotates round the ring in one direction or the other: the symmetry is given by a so-called 'twisted' cyclic group, which mixes both space and time. If you move one cell round the ring, *and* shift phase by a suitable fraction of a period, then the entire dynamics looks identical. The name indicates that the rotation round the ring is 'twisted' into the time direction as well, but the technical definition is more prosaic.

A ring of individual cells is called a *discrete* model: the components are separate lumps. Classical mechanics also – indeed more often – works with *continuum* models, in which the system is distributed in a continuous fashion. A line ——— is a continuum, whereas a line of individual dots ...... is discrete. Similarly a plane is a continuum, whereas a lattice of dots is discrete. If you imagine the dots getting very very close together, though, then the discrete model becomes a very good approximation to a continuum. This process is known as 'forming the continuum limit'.

Turing argues that the same two types of wave – stationary and travelling – can occur for the continuum limit of infinitely many infinitely thin cells, that is, a ring of tissue with circular symmetry. Again, he has found just some of the possible solutions, but perhaps the commonest ones. It's a moot point, whether to model a ring of biological tissue as containing a large finite number of cells or as a continuum, since both models produce similar predictions.

The stationary waves (another term is 'standing waves', which we use elsewhere) are the most interesting from the biological point of view, because it's easy to think of living creatures that have the corresponding form. Although circular rings of tissue are hard to come by, tissue with circular symmetry is not uncommon – for instance, the stalk of a plant, or perhaps segments of gut. We've already argued that many phenomena of symmetry-breaking are universal; and in fact the solutions that Turing finds are likely to occur in any circularly symmetric system.

The type of symmetry-breaking that is predicted by Turing's theory is manifested in a creature that begins by having circular symmetry but then develops some particular feature (triggered by morphogens) in equally spaced clumps around the circle. The branches of a plant or tree are a bit like this: a stalk with circular cross-section produces buds at its tip, which end up equally spaced along it. However, translational symmetries along the stalks are also involved in this case, so while it's the same kind of phenomenon, it's not the best example. The petals of a flower, growing on the end of a stalk with circular cross-section, might seem closer to what's needed; but Turing alludes to the curious predominance of particular numbers of petals as evidence that something more subtle must be going on. (Again, we might imagine that genetic switching, rather than reaction-diffusion dynamics, is selecting specific numbers. None of this explains why the genes prefer Fibonacci numbers!)

Turing's favoured example is the *Hydra* (Figure 7.8). This tiny creature lives in freshwater ponds, and is a bit like a miniature sea anemone. It has a bulgy cylindrical body with tentacles at one end. Turing describes an experiment in *Hydra* development:

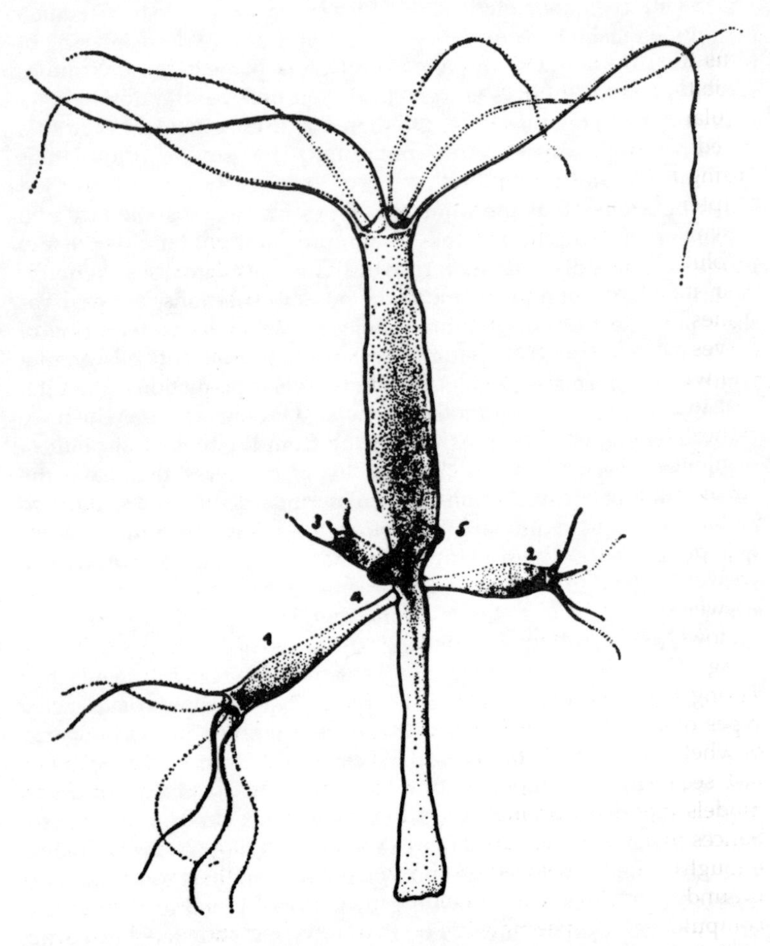

Figure 7.8 The hydra. This one has buds and is less than perfectly symmetric, but real hydras tend to be like that

> A part of *Hydra* cut off from the rest will rearrange itself so as to form a complete new organism. At one stage of this proceeding the organism has reached the form of a tube open at the head end and closed at the other end. The external diameter is somewhat greater at the head end than over the rest of the tube. The whole still has circular symmetry. At a somewhat later stage the symmetry has gone to the extent that an appropriate stain will bring out a number of patches on the widened head end. These patches arise at the points where the tentacles are subsequently to appear.

Thus the development of the *Hydra*, in this laboratory experiment, exhibits symmetry-breaking from circular symmetry $O(2)$ to that of a regular polygon – a dihedral group. The six-tentacled hydra illustrated has symmetry group D_6. All this is exactly in accordance with Turing's theory, and parallels what happens to the drop of milk in chapter 1, which loses symmetry from $O(2)$ to D_{24}.

Symmetry-breaking and chaos are part of an ongoing dynamical revolution: the replacement of classical 'linear' mathematics by 'nonlinear' mathematics. Linear models are based upon straight lines, flat planes, and constant growth rates. Nonlinear models are based upon curves and variable growth rates. If, like the classical mathematicians, you want to write down *formulas* that solve your equations, then it's best to stick to linear models. But, with computers and new mathematical tools such as topology, formulas have taken a back seat. Nonlinear models are more interesting, and generally fit nature better. All too often a nonlinear system has been forced into a linear mould (a round peg into a square hole!) in order to obtain an answer. The philosophy behind this approach seems to be that a wrong answer, or the right answer to the wrong question, is better than no answer at all.

However, linear models don't *always* give wrong answers. Considering when the work was done, it's not surprising to find that Turing's models are all linear. As a result, while he shows that certain types of symmetry-breaking can occur, he doesn't address the issue of whether they are to be typical. Turing is well aware of this, and the last section of his paper emphasizes the need to go to nonlinear models (because, among other things, in linear models the disturbances that create the pattern grow indefinitely large if you wait long enough). And his beloved computers finally edge their way in, on the grounds that nonlinear equations are too hard to solve without the computer. He is pessimistic about non-computational approaches: 'One cannot hope to have any very embracing theory of such processes.' Unduly pessimistic! For this book is about just such a theory; indeed one that substantially confirms his general points about symmetry-breaking.

The biologist Brian Goodwin has devised nonlinear computer models for development which also exhibit symmetry-breaking, but do not involve Turing's reaction-diffusion mechanism. In Goodwin's approach, the final creature is not so much a state of the system, but a kind of bifurcation diagram! The idea is that, as the creature develops, its older parts are 'frozen', becoming relatively fixed in form. Imagine a blob of thin jelly growing, but thickening up and hardening as it grows. Many plants grow in much this way – although it must be admitted that even a very ancient tree still adds a thin layer of bark each year. However, the growth at the new buds is much faster and more noticeable. If a creature grows in this way, then the age of each piece of tissue is like a bifurcation parameter. If the creature begins having circular symmetry, then the initial stages of growth have circular symmetry; and that symmetry is 'frozen' as the form becomes fixed by age. New growth continues to have circular symmetry, until a bifurcation occurs, say, via one of Turing's standing waves. So the creature develops a finite number of equally spaced patches of morphogen, which can trigger bulges, which grow into branches or tentacles or petals. So the symmetry of the creature changes as you cast your eyes along the tissue from old to new. If the direction of growth happens to be up, which is usual, then, you get a circular stalk with a lot of equally spaced projections at the top. The *Hydra* is just like this. So is the milk-splash.

More specifically, Figure 7.9 shows Goodwin's computer model of one step in the morphogenesis of *Acetabularia*, a single-celled marine alga. This creature begins as a spherical zygote, which puts out a root-like structure and a stalk. The stalk grows, and produces a ring of small hairs, called a *whorl*. The tip continues growing from the centre of the whorl, producing more whorls; then it develops a capped structure which gives it its common name, mermaid's cap. The production of the whorl is related to the calcium concentration at the tip. In the mathematical model, the calcium concentration at first has circular symmetry, but then the symmetry breaks. The graph of calcium concentration develops a series of ripples (Figure 7.9), just like Turing's stationary waves, which are precursors of the whorls. In the case illustrated, the symmetry breaks from $O(2)$ to D_6, although other numbers than 6 can also occur. The important point is that in Goodwin's model we observe a qualitative morphogenetic change occurring for purely dynamical reasons.

An amusing postscript to this idea is that developing projections might sometimes have roughly circular cross-section (say for reasons

Figure 7.9 Computer simulation of symmetry-breaking in the morphogenesis of the alga Acetabularia. *The graph shows ripples in the calcium concentration at the tip of the growing organism*

of chemical pressure-balance). If so, each individual projection may produce its own system of new sub-projections, by a second symmetry-breaking bifurcation, as in Figure 7.10. This kind of ca-

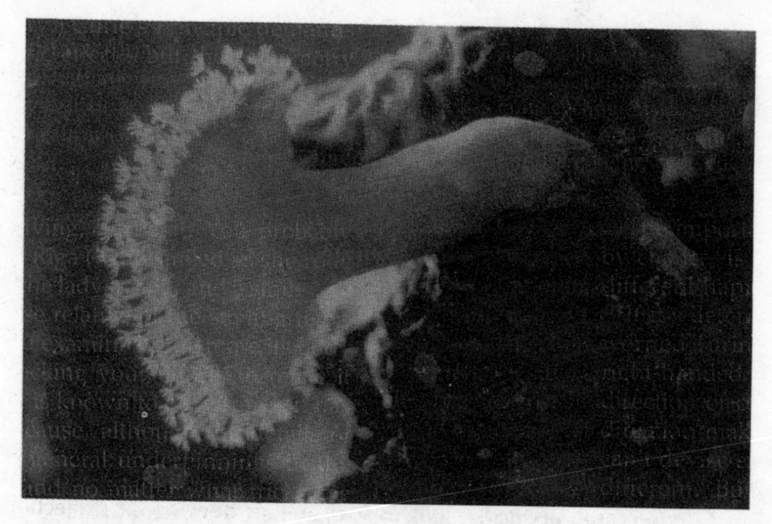

Figure 7.10 Polyps whose tentacles are polyps with tentacles

scade effect may explain why tree branches divide not just once, but repeatedly.

Gastrulation

Turing's final example, already mentioned, is gastrulation: the collapse of the spherical blastula to form a circularly symmetric tube (Figure 7.3, stages 10–12). The initial symmetry of this system, of course, is spherical! We've already asked whether this collapse might be due not to chemical influences, but mechanical ones, like shell-buckling – driven, however, by chemically induced pressure differences; and we've hinted that this isn't actually so. Turing considers the possibility that a morphogen, rather than mechanics, is responsible; and he comes to the same general conclusions about symmetry-breaking. Namely, the most likely pattern to develop is one with circular symmetry about some (random) axis, just as for mechanical buckling. Which theory is right? The universality of the phenomenon under investigation means that we can't distinguish between rival theories merely on the grounds of the patterns that result, so more detailed models and more specific experiments are needed to find out. We'll see in the next section that despite predicting the right symmetries, neither theory accords with all the facts!

The Missing Theory

Before getting too carried away with Turing's theories, it's worth taking a look at what is now known about biological development. The answer is: experimentally, a great deal; theoretically, very little. The biologist Wallace Arthur talks of the 'missing theory' of development.

We'll discuss some experiments first, followed by a theoretical interpretation. Figure 7.11 shows gastrulation in the echinoderm larva. The roughly spherical blastula does indeed collapse, forming a roughly circular depression. However, cells don't just deform mechanically: they actually come apart and migrate into the interior. The reason for this, however, is basically mechanical: the cells just change their degree of stickiness, come unstuck from their original positions, move around, and stick fast elsewhere. Is the precise motion of each cell programmed in the creature's DNA, or is just the general instruction 'migrate inside, look for somewhere sticky, and adhere to it'? Experiments show that the second statement hits the mark. What

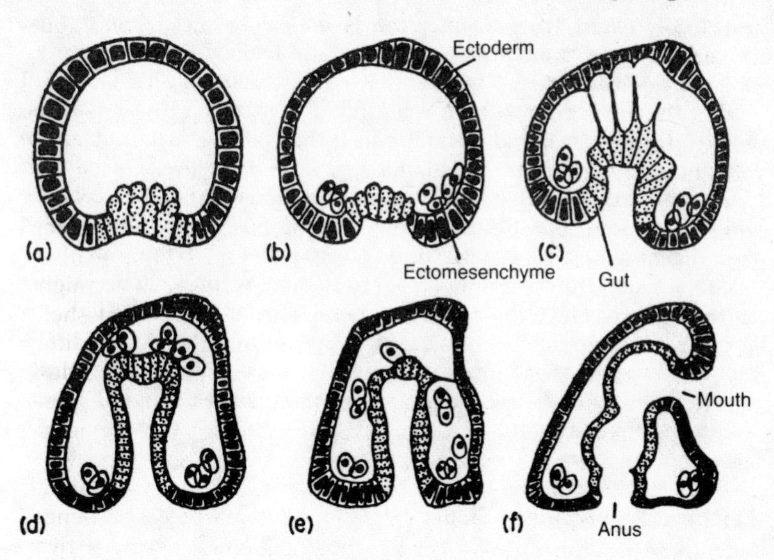

Figure 7.11 Gastrulation in the echinoderm larva

we observe is a 'free-running' piece of dynamics, although it's triggered in some way, either by mechanics or chemistry or instructions in the DNA, and it is also subsequently controlled by the DNA.

It should also be said that the axis along which the collapse occurs is not as random as Turing's theory would have it. The biologist Jack Cohen describes the true situation like this:

> The way in which the axes of the embryo form within the egg is determined by the position of the egg in the ovary, because of the necessary asymmetries of yolk, micropyles, and so on. There are a few eggs in which the position of sperm entry determines the anterior-posterior orientation of the future embryo, but the dorso-ventral axis is nearly always determined by the animal and vegetal poles of the egg, and hence by original yolk asymmetry in most cases. So the egg has acquired an architecture in the ovary which determines the whole symmetry of early development.

The general gist of this statement should be clear enough, even if you don't know what a micropyle is. (It's a pore through which sperm can enter.) 'Anterior-posterior' just means 'front-back', and the 'dorso-ventral axis' is the line along which the backbone and stomach develop – that is, the symmetry axis of the collapse that we are discussing. The 'animal pole' is where the main development goes on, and the 'vegetal pole' is usually where the yolk collects. The point is that the apparently spherical blastula in fact has a preferred axis, between the animal and vegetal poles; the collapse occurs near the vegetal pole. None of this invalidates Turing's basic line of reasoning: what it does is show that the small perturbations that initiate collapse are not random, but are determined by the positions of the animal and vegetal poles. In a sense, there is a 'weakness' in the wall of the blastula, like a thin region in a spherical shell; so the collapse occurs at that point. Nevertheless the general nature of the collapse is circularly symmetric, as Turing claims. So Turing's theory is at best a metaphor, perhaps appropriate to the analysis of changes in symmetry; it's not a complete description of the detailed biological effects that bring them about.

Left- and Right-handed Snails

While on the subject of actual biology, Turing raises a question in his paper about left- and right-handed phenomena in development, the answer to which was not known at the time, but which is now fairly clear.

Suppose, for the sake of argument, that we take on faith Turing's point that symmetry-breaking can (and often will) occur. In fact such faith recently acquired a more solid foundation. A group of scientists at the University of Bordeaux – Vincent Castets, Etienne Dulos, Jacques Boissonade, and Patrick De Kepper – have found the first *experimental* evidence for a Turing structure in a chemical reactor, nearly forty years after Turing's theoretical prediction that such patterns can occur. Then a difficulty arises, precisely *because* of the Extended Curie Principle: solutions come in symmetrically related bunches. Consider a creature, such as a snail, that develops a spiral shell. There are two types of spiral, left- or right-handed: they coil in opposite directions. The snail is said to be chiral: it has a definite 'handedness' – the technical term is *chirality*. The alien T'pots and Top'ts of Figure 7.1 are chiral. Suppose that the snail embryo begins from a spherically symmetric state (it's best to think of the first cell rather than the blastula, because the snail blastula is not very spherical!). Turing's argument that symmetry-breaking bifurcations can occur explains why it is possible for a spiral to form. But because it attributes broken symmetries to random disturbances at the onset of instability of the fully symmetric form, his theory leads inexorably to the conclusion that there should, on average, be roughly equal numbers of left- and right-handed spiral shells in any species of snails. For any random disturbance that leads to a left-handed snail, there is another equally probable random disturbance, its 'mirror image', that produces a right-handed snail!

Unfortunately, in most animal species, handedness is not equally distributed, so Turing's theory runs into trouble. The human heart, for instance, is almost always on the left; and most people are right-handed. Humans seem to be the only apes that prefer one hand over the other; but in several species of ape, us included, mothers cradle babies mainly – but not exclusively – on the left side. Mouse embryos can develop in left- and right-handed forms. And snail shells nearly always coil to the right when viewed from above. Our planet is like one in which alien T'pots exist, but their mirror-images, the Top'ts, don't. Or one where coins always land 'tails'. That's a very strange planet. So Turing's theory seems to be in trouble, even before its details have been worked out.

This problem doesn't arise for the choice of axis during gastrulation, because all that a different choice of axis does is to alter the *orientation* of the developing embryo. We don't consider the difference between a snail that happens to be developing head-down and one that develops head-up as being especially significant. When the fully developed snail starts to crawl around on the ground, both types will end up looking identical. So we don't get worried about the proportion of 'head-ups' compared to 'head-downs': indeed it never occurs to us to look for data on what that proportion is! But because no rotation in space can make something left-handed into something right- handed, the breaking of bilateral symmetry is different.

In search of weak points, Turing analyses the argument that there should be equal numbers of left- and right-handed creatures, showing that it requires three assumptions:

1. The laws of physics, as they apply to the chemical reactions and diffusions that are involved, are left-right symmetric.
2. The initial totality of zygotes (fertilized eggs) for the species is left-right symmetric.
3. The range of possible random disturbances is left-right symmetric.

Therefore, says Turing, in practice at least, one of these three assumptions must be wrong. But which? It's quite fascinating to watch him wriggling on this hook of his own devising.

He offers several suggestions, none entirely satisfactory. The first is that chemical molecules can occur in both left- and right-handed forms. Suppose that the distribution of such molecules in the morphogen is not fifty-fifty. Then in a sense all three of (1, 2, 3) are violated! In particular, the standard equations for the process of diffusion are left-right symmetric, but those take no account of the shape of the molecule that's diffusing. Turing also aruges that more accurate equations, incorporating the effects of the shape, would presumably make (1) invalid, because left-handed molecules wouldn't have the same effects as right-handed ones. (The equations for the left-handed case would look like those for the right-handed case, but reflected.) Or, he suggests, (2) might fail because of some asymmetry of the chromosome. (He's close, here; but the true source of asymmetry is the *ovary* of the mother snail, as we'll shortly see.) There's a valid general point, namely, a gross asymmetry of the developed animal can in principle be triggered by very tiny degree of asymmetry at the bifurcation point, so small differences of the kind Turing has in mind could suffice.

Turing isn't entirely happy with his own arguments, and in particular he says he would prefer something that works for models that depend only on concentrations of chemicals rather than their shapes. Given what's currently known about biochemistry, Turing's arguments are even less convincing, although some parts are heading along plausible directions. Few biologists believe that chirality in animals has anything much to do with chirality in chemical molecules. Left- and right-handed snails don't use left- and right-handed DNA: they use the same kind of DNA. The main point is that morphogens themselves – if they exist at all – will either be inherited from the parent in the germ cell that gives rise to the developing embryo, or assembled according to genetically programmed instructions. A better idea is that a left-right bias of morphogens could be propagated from parent to offspring, independently of chemistry.

Generation Gap

The correct answer, however, is that the genetics select for a final form with left-right bias: that is, that the bias is caused not by the 'free-running' chemical dynamics, but by the genetic switching that every so often interferes with it. Precisely because only tiny biases are needed, it's easy for the genetics to introduce such a bias. The full story is rather curious, and shows that the image of DNA as a 'program' for development is as naive as a literal version of Turing's theory is.

It is said that in 1894 a biologist by the name of Crampton showed that the direction of rotation of the snail's shell is determined very early on in cell-division, namely at the third cleavage, when the number of cells increases from four to eight! (In passing we note that, given the tetrahedral symmetry of the second cleavage, the four-cell stage, this is the earliest stage at which left-right symmetry of the arrangement in cells can be broken. So nature wastes no time in getting on with the job. However, there's a cryptic asymmetry earlier, visible even at the 1–2 and 2–4 cell divisions.) These eight cells form two layers, each with four cells; and the cells in one layer are smaller. The layer of smaller cells is rotated relative to the larger cells (Figure 7.12). If it rotates to the left, then a left-handed (sinistral) adult results; if to the right, then a right-handed (dextral) one.

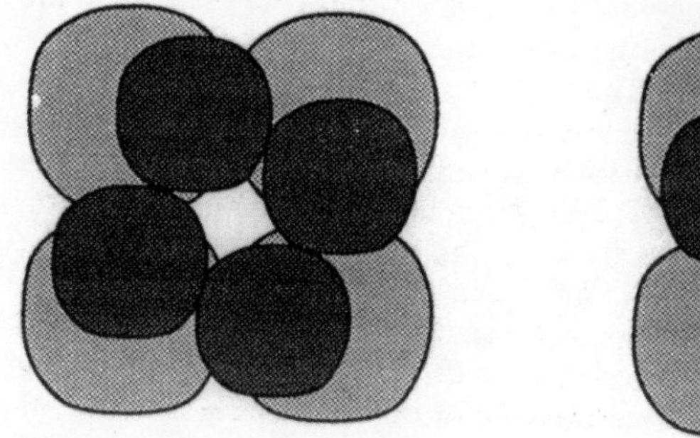

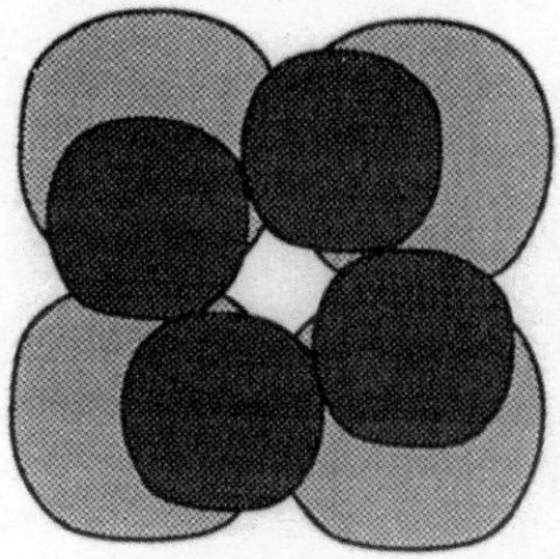

Figure 7.12 Early stages in the development of left- and right-handed snails

The important question now is: is this initial direction determined by chance, is it programmed genetically, or does something entirely different happen?

If it's determined by chance, then we're back in the situation that worried Turing, and we'd expect to obtain equal numbers of left- and right-handed snails. However, in most species of snails, only one direction of coiling occurs. Paradoxically, this very constancy of the direction makes it hard to find out why it's happening, because we can't devise experiments to see what might have happened if it were different. But a few species such as *Lymnaea peregra* and *Patura suturalis* exhibit both directions of coiling (Figure 7.13), and for these, experiments are possible: the genetics can be traced by crossing

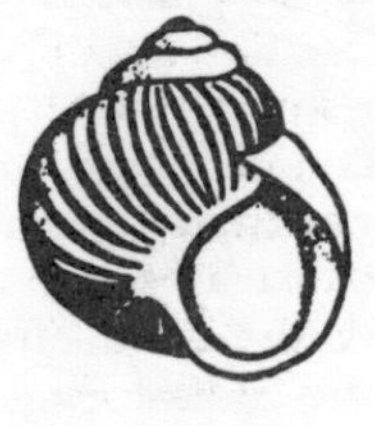

Figure 7.13 Left- and right-handed snails from the species Patura suturalis

different varieties and observing how many left-handed snails and how many right-handed snails you get. In 1930 A. E. Boycott, C. Diver, S. L. Garstang, and F. M. Turner, building on earlier work of A. H. Sturtevant, Boycott, and Diver, performed an exhaustive cross-breeding programme on about a million snails. They wrote it up as a paper called 'The inheritance of sinistrality in *Limnaea peregra*'. Here's what they found. The direction of coiling is genetically programmed; but in the genes of the *mother* rather than of the offspring! The reason is that the first few stages of cell division are controlled not by the DNA in the egg, but by its cytoplasm (cell-tissue), and the cytoplasm is passed on directly from the mother. So everything happens one generation later than you'd expect: what is inherited isn't a specific handedness, but the tendency to bear children with a specific handedness.

This explanation shows that the root of the problem is in Turing's assumption (2): it is indeed the case that the population of zygotes is biased towards a specific handedness. The biasing mechanism is subtle and surprising, but it is genetic in origin and occurs during the 'non-dynamic' part of development, to which Turing's theory is not applicable. So, interesting as it is, we won't consider it further.

Weak Interaction in the Primeval Oceans

None of the above fully explains how the asymmetry between left- and right-handedness in humans, or in the coils of snail shells, *originated*. It just explains how asymmetry can be *propagated*, once it has arisen. Clearly any asymmetry at a molecular level might show up in amplified form in a living creature, so the problem really resides at the molecular level. Amino acids, the basic molecular constituent of proteins, can come in both left- and right-handed forms; but typically one form is dominant and the other hardly ever occurs. Why?

One intriguing theory of the prevalence of particular handedness in biological molecules tracks it back to an asymmetry at the level of fundamental particles. Physicists distinguish four basic forces in Nature: gravity, electromagnetism, and the strong and weak interactions, about which we'll say more in chapter 10. The last two make their presence felt only at short range and on the scale of elementary particles. In 1956 T. D. Lee and C. N. Yang advanced the theory that the weak interaction violates chiral (left-right) symmetry: if the universe were reflected in a mirror, the weak interaction would not behave in exactly the same way that it does. As Wolfgang Pauli put it, 'the Lord is a weak left-hander'. Few physicists believed they were right: Pauli prefaced his remark with 'I do *not* believe that' and offered to 'bet a very high sum' – but they were, experimental evidence for this violation of symmetry in the decay of cobalt-60 being obtained by C. S. Wu in 1957. In the same year Lee and Yang were awarded the Nobel Prize. Pauli quickly revised his opinion: 'I am shocked not so much by the fact that the Lord prefers the left hand as by the fact that he still appears to be left-right symmetric when he expresses himself strongly. In short, the actual problem seems to be the question: Why are strong interactions right-and-left symmetric?' It's now well established that there's a fundamental violation of left-right symmetry in particle physics. But can an asymmetry which only affects elementary particles, and which is in any case very very small, really show up in which hand we use to write with?

Quite possibly. Dilip Kondepudi says:

> One can envisage a chiral-symmetry-breaking transition in the primordial oceans. Such a transition can take place if there is a suitable chiral auto-catalysis and if the system is driven far from thermal equilibrium due to an inflow of reactants into the primordial oceans. In this scenario ... slow passage through the transition point will correspond to slow increase of those concentrations. If there were no systematic biases, then we must expect the dominance of left- and right-handed molecules to occur with equal probability. Hence on the surface of the Earth we must expect to have regions dominated by one kind or the other – somewhat like domains in a ferro-magnet. On the other hand, if there was a systematic bias, and if there was a slow growth in the concentration of reactants, then our theory tells us that the dominance of the favoured handedness can occur with high probability even if the bias is very small. Consequently only the favoured handedness will dominate over the entire planet. As a possible source of bias, parity violation in weak interaction has long been of much interest, ever since

> its discovery. However, most of the previous considerations were of the opinion that the chemical effects of weak interactions are too small to be of significance. We have investigated the question from the viewpoint of sensitivity due to slow passage through the transition point. The results show that if the production rates of left- and right-handed molecules differ by only one part in 10^{17} under prebiotic conditions, a slow growth of concentration on a time scale of 10^4–10^5 years could result in the dominance of the favoured handedness with a probability larger than 98%.

A key constituent of living creatures is proteins; and proteins are very large molecules, made from much smaller molecules known as amino acids. Remarkably, one part in 10^{17} is precisely the order of magnitude of the difference in energies computed by S. F. Mason and G. E. Tranter between left- and right-handed versions of an amino acid. Moreover, the naturally dominant amino acid has the lower energy. So it may be that nature repeatedly amplifies the tiny asymmetry of the weak interaction between elementary particles – first in the formation of the molecules of amino acids, then in the proteins that they form, then in the process by which those proteins control the development of an embryo, and finally in the growth and development of the embryo into an adult creature. A weak left-handed God creates left-handed amino acids that build left-handed proteins that form a left-handed embryo that grows into a left-handed adult. (We are not here referring to which hand the adult uses to write with, but merely to the propagation of asymmetry.) Curie's Principle demands an asymmetric cause for an asymmetric effect – but the asymmetry can be *very* small indeed.

The Grin on the Tail of the Tiger

That section title will take some justifying, but wait. You probably recall the sad fate of the young lady of Riga ('who smiled as she rode on a tiger. They finished the ride with the lady inside, and a smile ...') Turing's theory is reminiscent of a close relative of the smiling tiger, namely the Cheshire cat: the closer you examine it, the more its body vanishes ... but the grin remains, mocking your efforts to make it disappear altogether. For a theory that is known to be 'wrong', it has an odd resilience. Probably this is because, although the details of Turing's theory don't hold water, its general underpinnings in the breaking of symmetry are universal, and no matter what tricks the genetic programs play, they are fundamentally opportunist in nature and tend to build upon universal dynamic phenomena. The tale of the tiger – more accurately, the *tail* of the tiger, and of the lion, the leopard, and the cheetah – is a remarkable case in point.

To a first approximation, tigers are cylindrical. In chapter 5, on Couette-Taylor flow, we've seen one possible way for cylindrical symmetry to break: to spirals. Another type of symmetry-breaking occurs when the translational symmetry is broken: then a uniform pattern is replaced by periodic stripes. William Blake, as we've said, was impressed by the tiger's 'fearful symmetry'. But not just bilateral, oh no. Did Blake stumble on a deeper truth than he imagined?

Suppose, following Turing, that the patterns on a mature tiger's skin are controlled by chemicals which diffuse over the surface of an embryonic tiger as it grows. If we model the tiger by a perfect cylinder then a fully symmetrical pattern of pigmentation would give just a uniform orange tiger, in other words a lion. But a uniform distribution of chemicals can be unstable. Then the symmetry breaks – and one possibility is stripes! Is this the main difference between lions and tigers?

Jim Murray of Oxford University rather thinks so, and he has used a computer to model the same phenomenon on a more realistic approximation to a whole animal. He finds patterns (Figure 7.14) that

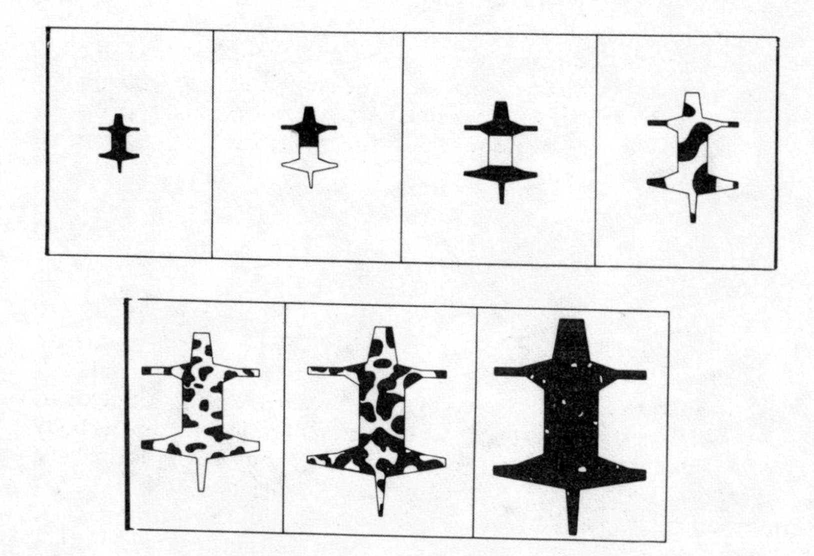

Figure 7.14 Simulation of patterns on a bilaterally symmetric (and almost front/back symmetric) animal skin. From the fourth picture onwards, left-right symmetry is broken

break the animal's left-right symmetry; moreover, as Turing noted, any non-uniform pattern breaks the local symmetry of the animal tissue. Turing himself obtained 'dappled' patterns similar to those on some breeds of cow.

Even more fascinating are Murray's models of animals' tails. Tails are a lot closer to cylinders than complete animals are, and the patterns actually found on the tails of mature cats closely resemble those obtained by computer simulations of reaction-diffusion equations. For greater accuracy still, the cylinder may be replaced by a slowly tapering cone, to correspond to the gentle taper of a real tail. Now the resemblance is even stronger (Figure 7.15); moreover, we can interpret the tapering tail as a sequence of circular cross-sections, forming a kind of 'frozen' bifurcation diagram of the kind we discussed above when talking about the *Hydra*. Then we can see the onset of the instability of stripes: tails often have striped ends, but the stripes break up into spots as the tail becomes fatter.

The idea can be pushed further. In Couette–Taylor flow the 'striped' pattern of Taylor vortices breaks symmetry again, to wavy vortices. In an animal analogy, the stripes would break up into spots, arranged – like wavy vortices – in a roughly hexagonal pattern. Now

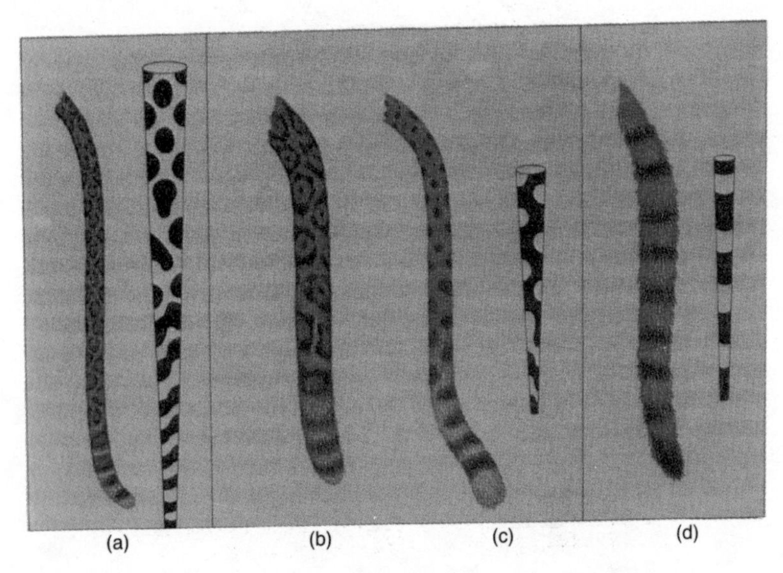

Figure 7.15 Circular symmetry-breaking bifurcation in tapered tails, in nature and in computer simulation (a) Leopard. (b) Jaguar. (c) Cheetah. (d) Genet

the tiger has become a leopard! It's said that leopards can't change their spots, but the spots may have arisen when tigers changed their stripes.

One of Turing's results is that, roughly speaking, there's a natural scale for the separations between stripes in patterns produced by reaction-diffusion systems. This means that long thin stripes are less stable than short fat ones, so thin enough stripes will break up into spots. Now an animal's tail is thinner than its body, so stripes on the tail are shorter than those on the body. They thus break up less easily. So a spotted animal can have a striped tail but a striped animal cannot have a spotted tail. This qualitative 'postdiction' by Murray is borne

Figure 7.16 Imaginary animals whose markings could in principle be formed by symmetry-breaking. (a) Spigers. (b) Squeopards

out across the animal kingdom, a striking triumph for this kind of thinking. However, presumably only instability prevents the occurrence of spigers – tigers with spiral stripes (Figure 7.16a), or squeopards – leopards whose spots are arranged in squares rather than hexagons (Figure 7.16b).

When thinking about the markings on animals, it's important to realize that they're 'laid down' fairly early in the creature's development, when its shape is very different from its final form. Jonathan Bard has studied three species of zebra, whose adult forms have quite different markings, and discovered that at a particular embryonic stage the precursors of those patterns are identical. They just grow differently – basically, it's a matter of *timing*. Another example is the spots on a peacock's tail, which *look* very like something that might be produced by diffusion. Unfortunately for this theory, when those spots are laid down, the feathers are wound up into cylinders! So there's a lot left for us to understand.

Not Too Literally ...

Remarkable as the resemblances between simulation and real animal may be, it's important to realize that we haven't established the literal truth of Turing's theory. Indeed, as we've repeatedly argued, the basic phenomena of symmetry-breaking are *model-independent*. They don't depend upon specific mechanisms assumed in the mathematical model, but only upon its general form: specifically, upon its symmetries. This implies that similar patterns might arise by quite different physical, chemical, or biological mechanisms. It's a tautology: model-independent results cannot be used to confirm specific models! Turing didn't have the luxury of a general theory of model-independent dynamics, and his reaction-diffusion equations can profitably be viewed as a simple test-bed for more general ideas; but there's no reason to suppose that animal patterns are really determined by nothing more complex than diffusing morphogens.

Indeed there are plenty of reasons not to suppose this, and we've explained some of them in passing. Here's another, revealed by making a closer examination of the patterns on real animal skins. The light and dark pigment that we see is generally concentrated at the tips of the animal's hair. Does the morphogen have to diffuse right up the hair? That would cause all sorts of difficulties for the theory, because the hairs grow and then fall out. Fortunately not. It's not unreasonable that morphogens in the skin (or an earlier stage of embryonic development) can influence the base of each hair, and thus the colour at its tip. Turing breathes again.

Prematurely. In some animals, the hairs don't poke straight out, but lie flat. In the leopard (but not in the cheetah) hairs lie across several different regions of the complicated patterning of the spots! So the colour of the hair changes along its length, in just the right way to form the observed pattern on the animal's body. How does it *do* that? A very good question: nobody seems to have a definite answer! It's as if the developing cells 'know' where they're going.

There are several theories of how cells might possess this kind of 'positional information'. Louis Wolpert demonstrated one in his 'French flag model' (Figure 7.17), which shows that a single chemical gradient (a concentration that decreases regularly in some particular direction, from a 'source' to a 'sink') can act as a kind of map-reference. A gradient could affect the colours along a single hair – but how does the hair 'know' where it's going to lie down, so that all the

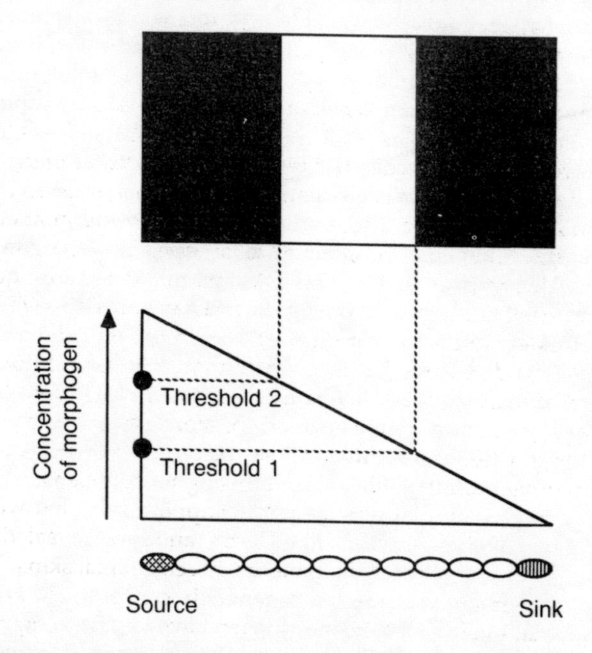

Figure 7.17 Wolpert's French flag model. A morphogen diffusing from source to sink creates a concentration gradient (sloping line). Thresholds of the concentration can trigger distinct processes (regions of the flag)

hairs fit together properly? It seems baffling – though it's possible that the question makes no more sense than asking why people's legs are just long enough to reach the ground. The effect is presumably a consequence of the general dynamics of hair-formation.

Again, we suggest taking an appropriately laid-back view. There's nothing sacred about the details of Turing's model. Historically the symmetry-breaking argument was embodied in Turing's model, but that was for convenience only: Turing lacked the concepts and viewpoint needed to state it in a model-independent way. Because of the universality of symmetry-breaking, it's likely that similar effects will occur in other models. Presumably the multicoloured hairs of the leopard, and their propensity to lie down in just the right place, are fine detail of some more sophisticated model. (Yes, it would be nice to know which!) We're still left with a strong feeling that somewhere along the line, genetics learned to take advantage of the natural patterns that are afforded by symmetry-breaking. It was perhaps a matter of evolutionary opportunism: why reinvent the wheel (rather, the stripe) when it's already available as a mathematical universal? There are strong reasons to believe that buried under the intricate biological details of morphogenesis there are basic mathematical patterns, and paramount among them are symmetry-breaking bifurcations.

Let's leave the final word to Charles Darwin, who ended his epoch-making book *The Origin of Species* with these words:

> There is grandeur in this view of life, with its several powers, having been originally breathed by the Creator into a few forms or one; and that, while this planet has gone cycling on according to the fixed law of gravity, from so simple a beginning endless forms most beautiful and most wonderful have been and are being evolved.

8

The Pattern of Tiny Feet

A centipede was happy quite,
Until a frog in fun
Said, 'Pray, which leg comes after which?'
This raised her mind to such a pitch,
She lay distracted in a ditch
Considering how to run.

Mrs Edmund Craster, *Centipede*

When you want an automobile to go faster, you don't always just put your foot on the accelerator pedal and go. Unless the car has an automatic transmission. If not, then every so often you have to change gear. The range of comfortable speeds for the engine isn't as wide as the range of car speeds that the driver may need to use, so a system of gears is employed to alter the relationship between engine speed and wheel speed. Horses also have a kind of gearing system: they don't always move in the same way. At low speeds they walk; at higher speeds they trot; and at top speed they gallop. Some insert yet another type of motion, the canter, between trot and gallop. Similarly, humans walk at moderate speeds, but run if they need to move quickly. The differences are fundamental: a trot isn't just a fast walk, but a different kind of movement altogether.

These distinct patterns of leg-movement are known as *gaits*, and their study, naturally, is called gait analysis. An elderly Professor of Biology, under the impression that the word 'gait' derives from something French, always used to lecture about 'gays', causing considerable confusion among his students. In fact the word is pronounced 'gate', and according to the *Oxford English Dictionary* has the same origins as the word for the thing you open to get at your garden path – though precisely what the *OED* means by that is less than crystal clear. In Barclay's *Ship of Folys* of 1509 we find 'their gate and looke proude and abhominable'. The spelling changed between the seventeenth and eighteenth centuries and 'gait' then became standard.

Most gaits possess a degree of symmetry, a fact first emphasized by M. Hildebrand around 1965. For example when an animal, such as the long-tailed Siberian souslik, bounds, both front legs move together and both back legs move together (Figure 8.1). This gait preserves the bilateral symmetry of the animal. Other symmetries are more subtle: for example the left half of a camel may follow the same

Figure 8.1 The bound of the long-tailed Siberian souslik retains bilateral symmetry

sequence of movements as the right half, but half a period out of phase – that is, after a time delay equal to half the period (Figure 8.2). This is yet another example of symmetry-breaking: here the gait of a bilaterally symmetric animal need not be bilaterally symmetric.

However, such a gait has its own characteristic symmetry: 'interchange left and right sides and shift phase by half a period'. You use exactly this type of symmetry-breaking to move yourself around: despite your bilateral symmetry, you don't move both legs simultaneously! There's an obvious advantage to bipeds in not doing so: if they move both legs at once they fall over.

What controls gaits, and their selection at different speeds, and why do symmetries arise? It's a fascinating subject, which is why we've given it a chapter of its own. It's also a biological example of symmetry-breaking that doesn't have the same difficulties of interpretation that we've just encountered in Turing's theory. And it's a striking thought that everybody carries an example of symmetry-breaking around with them; or rather, they carry themselves around *on it*.

Moreover, it has technological significance. Some tasks are not suitable for wheeled vehicles – surveying the surface of Mars, or decommissioning a nuclear power station, for instance. Many engineers think that we should take a cue from nature and make walking vehicles. The control of the walking gait, and by extension of

Figure 8.2 The pace of the camel breaks bilateral symmetry

other gaits, then becomes a serious practical problem. It's also a surprisingly difficult one. This leads to some scientific curiosities, among them the robot pogo-stick. At first blush this looks like a candidate for Senator Proxmire's golden fleece award (for pointless science), but the point is that a one-legged gait is the simplest of them all, and the only way to obtain it is to hop, hence the pogo-stick. Having learned to control a single pogo-stick, one may then move on to the trickier problem of four coupled pogo-sticks, a mechanical quadruped. Lots of things make sense when you appreciate that science works in simple steps – including some of the work blessed by the attentions of the good senator.

What lies behind the patterns and changes of animal gaits? There are many clues in the literature. R. B McGhee and A. K. Jain remark that 'animals typically employ their limbs in a number of distinct periodic modes', a phrase that strikes an immediate chord for anyone familiar with dynamical systems. The main thrust of this chapter is to describe a striking analogy between the typical patterns of periodic oscillations in symmetric dynamical systems, and the different gaits employed in animal locomotion. What are symmetric gaits and how can they be classified? What is the evidence for and against nonlinear effects and symmetry-breaking in animal locomotion? What conclusions can be drawn about the general nature of locomotion, without considering specific model equations for the dynamics?

No Bounds

Interest in the patterns of animal gaits is probably as old as the human race itself. In his *De Incessu Animalium* the Greek philosopher Aristotle describes the walk of a horse:

> the back legs move diagonally in relation to the front legs; for after the right fore leg animals move the left hind leg, then the left fore leg, and after it the right hind leg.

However, he thought that the bound is impossible:

> If they moved the fore legs at the same time and first, their progression would be interrupted or they would even stumble forward. ... For this reason, then, animals do not move separately with their front and back legs.

After Aristotle, the next substantial advance in the classification of gaits and the understanding of the principles behind them was made by Fabricius ab Aquapendente in his *De Motu Locali Animalium* of 1618. We've already mentioned Fabricius in the context of the rotation of the sun: he was an intellectual rival of Galileo, and died of poisoning. The science historian Julian Jynes says that 'Coming from [Aristotle's treatises] to Fabricius' work is like emerging from a teeming, disorderly, and exciting town into a neat meadow, a more coherent panorama of observations.' Fabricius organized his ideas into neat categories: creeping, flying, swimming, walking in bipeds, walking in quadrupeds and multipeds. He noted that the type of locomotion used depends upon the terrain, and he tried to derive all locomotion as a combination of two basic types: walking with diagonal limbs in unison and leaping with opposite limbs in unison. We'll see below that these correspond roughly to the 'in-phase' and 'out-of-phase' oscillation patterns, so Fabricius had taken at least one step along the track that leads to symmetry-breaking. Aristotle's basic philosophy was that everything in nature has some function; Fabricius restated this as a principle of economy, that Nature avoids waste. Aristotle maintained that Nature does whatever is best, whereas Fabricius – in an astonishing anticipation of Darwin – stated that Nature *perpetuates* what is best.

William Harvey, a student of Fabricius best remembered for his discovery of the circulation of the blood, emphasized the role of the brain in locomotion, comparing it to a choirmaster, with an exquisite sense of rhythm and harmony. To prove this he cut the head off a chicken, after which its behaviour became less organized than before. Here we have the germ of another key idea, that of rhythmic neural control of locomotion, which we shall shortly meet again under the term 'central pattern generator'. René Descartes took this idea further; he considered the mind and the body to be two parts of a single whole, the body being a machine driven by the mind, which at the time was a revolutionary thought. However, he mistakenly thought that the nerves acted rather like a hydraulic system, with liquids ('animal spirits') being pumped along them. This was disproved when Jan Swammerdam immersed a frog muscle in water, pinched a nerve to make the muscle contract, and showed that the water level didn't change. Therefore nothing was being added to the muscle to make it contract.

Mathematics entered with a vengeance in the *De Motu Animalium* of Giovanni Borelli in 1860. Borelli applied mechanics and geometry to animal motion, showing – for example – that bones act as levers and obey the correct laws of forces. While the savants of the seventeenth century failed to solve any of the basic problems of animal motion, they took a problem that had been buried amid mysticism and confusion, and sorted it into its component sciences.

Horse Cents

Observing just how an animal's legs move as it runs or gallops is impossible without advanced technological assistance, especially high-speed photography. According to legend, this technique of modern gait analysis originated with a bet on a horse. In the 1870s Leland Stanford, former governor of the state of California, had an argument with Frederick MacCrellish over the placement of the feet of a trotting horse. Stanford put $25,000 behind his belief that at times during the trot, the horse had all of its feet off the ground. To settle the wager a local photographer, who was born Edward Muggeridge but adopted the pretentious name Eadweard Muybridge, was asked to photograph the different phases of the gait of the horse. Stanford, it is said, won his bet. The tale may be apocryphal, but it's certainly true that Stanford eventually commissioned Muybridge to photograph the movement of other animals, including humans, engaged in various activities. Later in his career, Muybridge murdered a man who was having an affair with his wife; he was acquitted on the ground of temporary insanity. Muybridge pioneered the quantitative study of gaits, and adapted a mechanical device known as the zoetrope to display them as 'moving pictures'. He first named his modified gadget the zoographiscope, then changed it to zoogyroscope, zoopraxinoscope, and finally zoopraxiscope.

Most gaits can be represented as symmetric cyclical patterns of movement. Successive cycles are generally very similar, and for purposes of mathematical modelling they may be assumed identical, so a single cycle represents the entire pattern. By convention, one gait cycle is the time between one foot striking the ground and the same foot striking it again, although there are a few gaits for which this convention doesn't work so well. We'll return to this point later, but for the moment we'll keep the discussion simple.

An important numerical measure for gaits is the *duty factor* of each foot. This is the fraction of the gait cycle for which the foot touches the ground. When an elephant ambles, for example, any given foot stays on the ground about three quarters of the time. Thus the duty factor is 0.75. For all of the gaits that we describe, we assume that all feet have the same duty factor, although in practice there are slight variations between one foot and the next. We'll mainly discuss bipedal (two-footed) and quadrupedal (four-footed) gaits.

The quantity normally used to compare the states of different feet, and hence to describe the gait's qualitative form, is the *relative phase* of a foot. This is the fraction of the gait cycle between a reference foot (normally the left front foot in quadrupeds) hitting the ground, and the foot concerned hitting the ground. In the human walk, the right foot hits the ground half a period later than the left: it has a relative phase of 0.5 compared to the left foot. The reference foot, in consequence of the definition, always has a relative phase of zero. The relative phase is crucial for determining the symmetry of the gait, but the duty factor plays virtually no role in this. For instance, the human run and walk have the same relative phases, and the same symmetries, but different duty factors.

Bipedal Gaits

Most higher animals, including both bipeds and quadrupeds, are bilaterally symmetric – like us, they have a plane of reflectional symmetry. Some bipeds, for example birds, have only two limbs – unless you consider a wing as a limb – while others (such as us) use only two for locomotion, with the other two never touching the ground. This is a questionable distinction because limbs that don't touch the ground can still play a role in gaits: for example, when we walk we swing our arms. For simplicity we'll ignore limbs that don't hit the ground, which neatly classifies humans as bipeds.

There are two basic bipedal gaits. The two legs can be out of phase with each other, that is, doing the same thing but hitting the ground at different times. Walking and running are examples in humans. Alternatively, both legs can be in phase: they do the same thing at the same time. Two-legged hopping is an example of this. The hopping gait preserves the animals's bilateral symmetry, so it doesn't involve symmetry-breaking. Walking and running break the bilateral symmetry, in the following sense: if you watch a creature walking and simultaneously watch it in a mirror, then the two images don't do exactly the same thing. However, they do *almost* the same thing: when the real left leg hits the ground, what seems to be the mirror image's right leg hits the ground. So both gaits stay the same provided the creature is reflected about its plane of symmetry *and* the phase of the gait cycle is shifted by half a period. This kind of symmetry shouldn't surprise us: we saw in chapter 3 that the natural symmetries of periodic motion combine both space (reflections, rotations, etc.) and time (phase shifts). Nature seems to be aware of this theorem.

Although the human walk and run have identical symmetries, the duty factor distinguishes them easily. The duty factor for a walk is always greater than 0.5, whereas that for a run is less. Olympic walkers hover uneasily at the 0.5 duty factor boundary. The difference is qualitative in the sense that 0.5 is not an arbitrary figure: it marks the point at which the entire animal can be off the ground at some stage of the gait cycle. The duty factor and the symmetry capture different features of gaits.

Quadrupedal Gaits

With more legs to operate, there are more possibilities for gaits – the source of the centipede's dilemma. Here are the eight commonest quadrupedal gaits, together with a potted description. The phase relationships are summarized in Figure 8.9.

- *Walk* The legs move a quarter of a cycle out of turn (a 'rotating figure-8' wave, either front left/back right/front right/back left or the reverse). The walk is a statically stable gait – the animal's centre of gravity is always within the polygon defined by the supporting feet. The duty factor for a walk is greater than or equal to 0.75 for quadrupeds. The *amble* is the running form of this sequence and has a lower duty factor. Figure 8.3 shows an elephant ambling.

Figure 8.3 The amble of the elephant

- *Trot* Diagonal pairs of legs (left front/right back and right front/left back) move together and in phase. One diagonal pair is half a period out of phase with the other pair. Figure 8.4 shows the trot of a horse.

Figure 8.4 The trot of a horse

- *Pace* This uses left/right pairing. The left legs move together and in phase. The right legs move together, half a period out of phase with the left legs. We've seen the pace in Figure 8.2.
- *Bound* This uses front/back pairing. The front pair of legs move together and in phase. The back legs move together, half a period out of phase with the front pair. We've seen the bound in Figure 8.1.
- *Transverse gallop* This gait resembles the bound, but the feet of the front and back pairs are slightly out of phase with each other. The left back leg is half a period out of phase with the left front leg; the right back leg is half a period out of phase with the right front leg. Figure 8.5 shows the transverse gallop of a cheetah.

Figure 8.5 The transverse gallop of a cheetah

- *Rotary gallop* This is similar to transverse gallop, except that the left and right back legs have swapped patterns. Diagonal legs are half a period out of phase with one another. Figure 8.6 shows the rotary gallop of a horse.

Figure 8.6 The rotary gallop of a horse

- *Canter* The right front/left back legs move together and in phase. The left front/right back legs move half a period out of phase with one another, and a rather arbitrary amount out of phase with the strongly coupled diagonal pair. The entire gait may occur left/right reflected, so there's a 'left-footed' canter as well as a 'right-footed' one. Figure 8.7 shows a horse cantering.

Figure 8.7 The canter of a horse

As well as these, there's a rarer gait, but one that completes the role played by symmetry:

- *Pronk* All four legs move at the same time. The pronk is uncommon outside of cartoons (Figure 8.8). The first author's first cat, a beast rejoicing in the name of Seamus Android, was observed, as a kitten, to pronk when startled by a sudden noise. (It also fell off the roof of the garden shed into a tank of water, a gait with no obvious symmetry.) The pronk is sometimes seen in fast-moving deer.

Different animals prefer different gaits. At low speeds, the walk is common. Most quadrupedal mammals trot at low running speeds; however, camels generally pace. Elephants amble; indeed elephants are almost the only mammals that amble, and that's all they can do! With increasing speed, a horse will walk – trot – canter – gallop. The canter (which we'll later see is rather a strange gait) seldom comes naturally: most horses have to be trained to do it, and some can't manage it at all. Wildebeests change directly from a walk to a canter.

Figure 8.8 A feline pronk

Reptiles trot at running speeds, but some lizards can run bipedally on their hind legs. Young crocodiles, but not adults, have been known to gallop, both rotary and transverse. At galloping speeds, small mammals such as mice use the bound; slightly larger mammals such as squirrels use the *half bound*. In this gait, which isn't listed above, but is illustrated in Figure 8.13 below, the front legs are slightly out of phase, similar to the front legs in a gallop.

Central Pattern Generators

Quadrupedal locomotion is fiendishly complicated: it requires control of more than twenty joints, and many more muscles. What controls locomotion? How does an animal know when to switch gaits? The poem about the centipede at the head of this chapter suggests that it may be better if it *doesn't* know. Italo Svevo, in his *Confessions of Zeno*, describes a similar predicament:

> He told me with amusement that when one is walking rapidly each step takes no more than half a second and in that half second no fewer than fifty-four muscles are set in motion. I listened in bewilderment. I at once directed my attention to my legs and tried to discover the infernal machine. I thought I had succeeded in finding it. I could not of course distinguish all its fifty-four parts, but I discovered something terrifically complicated which seemed to get out of order directly I began thinking about it.

Centipede and Svevo have the same problem. Fortunately, the control of locomotion is generally a subconscious activity: while we can decide to run instead of walking, we run naturally and don't consciously decide which leg to move. Our eyes do keep the brain informed about terrain, and that can override the subconscious programming to some extent.

Recall that animals' brains control the creatures' movements by sending electrical signals along the neurons, or nerve cells – long thin cells that function like telephone lines. The nerves hook together a bit like the components of an electronic circuit, and indeed we speak of *neural circuitry*. Theorists have suggested that locomotion must be controlled by things called *Central Pattern Generators*, or CPGs. A CPG is a system of neurons somewhere in the central nervous system – not necessarily in the brain itself – that produces rhythmic behaviour.

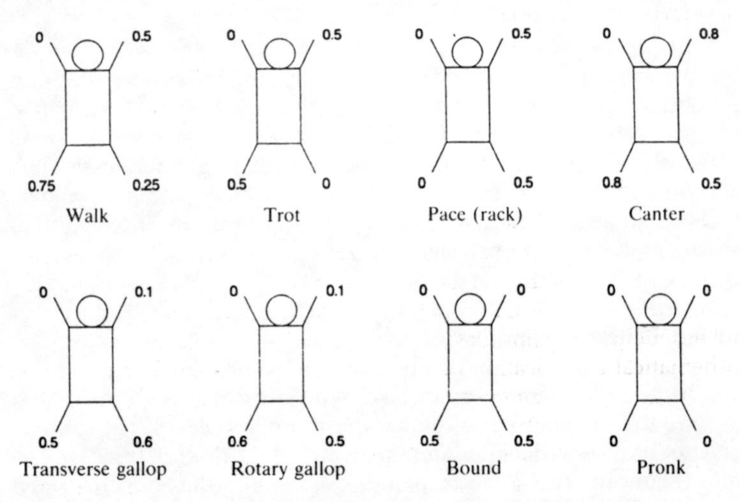

Figure 8.9 Phase relationships in gaits: the animal is a quadruped viewed from above

This isn't a new idea. It was arguably anticipated by Harvey. In 1879 T. H. Huxley compared the rhythmic motion of a crayfish to the melody played by a musical box.

> It is in the ganglia that we must look for the analogue of the musical box. A single impulse conveyed by a sensory nerve to a ganglion, may give rise to a single muscular contraction, but more commonly it originates in a series of such, combined to a definite end.

In some sense it's obvious that CPGs *must* exist: after all, locomotion does occur, and something in the brain or the nervous system must control it. There's a lot of experimental evidence in favour of the general idea. In 1935 E. von Holst showed experimentally that an assembly of neurons in the spinal cord can coordinate fine movements in some fish, without receiving any messages from the brain. However, the CPG theory is controversial, for at least two reasons. Firstly, the system of neurons that controls locomotion need not be all in one place: for example, each limb could have its own control circuit, with all four circuits in communication. This would be a Distributed Pattern Generator (DPG) rather than a central one. Some biologists seem to think that a DPG would still have to be driven by a CPG in order to provide the correct phase relationships for gaits; but on the contrary, we'll see below that nonlinear dynamics is quite capable of running a DPG without any centralized control, because symmetry-breaking occurs the same way in both CPGs and DPGs.

Symmetries of Gaits

The most striking general feature of animal gaits, and the reason for their presence in this book, is that they possess a great deal of symmetry. In fact, they possess more symmetry than is usually acknowledged by gait analysts, who refer to 'symmetric' and 'asymmetric' gaits when they actually mean 'highly symmetric' and 'less symmetric'. Recently G. N. Schöner, W. Y. Yiang, and Jack Kelso, and independently Jim Collins, have used group theory to give a mathematical classification of the symmetries of animal gaits. What do we mean by a symmetry of a gait? We distinguish two types. The first, spatial symmetry, refers to permutations of the oscillators (interpreted either as neuron systems in a CPG or as legs) in a coupled system. The second, temporal symmetry, involves patterns of phase-locking. The prime message from the mathematical theory of Hopf bifurcation with symmetry is that the symmetries of oscillating systems operate both in 'space' and in time. We've put quotation marks around 'space' to show that we're not just thinking of rigid motions in ordinary space: we're thinking of any non-temporal symmetries. The time symmetries, of course, are phase shifts. Now the symmetries observed in gaits do indeed specify a phase relationship between two or more legs: see Figure 8.9. They're precisely the mixtures of space (swap legs) and time (shift phase) that the mathematics leads us to anticipate.

The potted descriptions of the eight commonest gaits, above, describe their symmetries in precisely this fashion. To make the point crystal clear, let's discuss a few representative gaits in detail. The most symmetric gait of all is 'stand'. All four legs do the same thing (namely nothing at all!) and what they do doesn't depend upon time. The next most symmetric gait is the pronk, in which the animal 'stands' in a more energetic fashion. The pronk is completely symmetric in 'space' – all four legs do exactly the same thing – but it breaks the temporal symmetry of 'stand', which is a steady state, becoming time-periodic. That's extremely reminiscent of Hopf bifurcation! The stand is invariant under time-translation through an arbitrary interval; the pronk is invariant only under time-translations through integer multiples of the period.

The trot has fewer symmetries than the pronk, but they're more interesting. In a trot, the front right leg does exactly the same thing as the back left, so there's a spatial symmetry: 'interchange front right and back left'. There's another, 'interchange front left and back right'. Are these all? Well, you could interchange both pairs, of course; or leave all four legs in the same place, but those are consequences of the two symmetries just mentioned. There is, however, an additional symmetry. If we reflect the left and right sides of the animal, and shift phase by half a period, then the trot remains unchanged. The trot has a mixed spatio-temporal symmetry.

The pace and bound have very similar symmetries to the trot, but the pairings are different. Instead of elaborating all of the details, let's take a look at the rotary gallop, where another issue surfaces. It has some of the symmetries of the trot, but not all – namely, diagonal pairs of legs are half a period out of phase. However, no leg does exactly the same thing as any other leg. For example, there's a small phase difference between the front two legs.

We've now established two important general features of animal gaits. The first is that (as for all periodic motions of symmetric systems) their symmetries are mixtures of space and time. The second is that some gaits are less symmetric than others – transitions between gaits break symmetry.

Map of the Cat

The basic idea behind CPG or DPG models is that the rhythms and phase relations of animal gaits are determined by relatively simple neural circuits. What might such a circuit look like? There are many ways to approach this question. We might try to dissect out the actual neural circuitry in a real animal. We might design mathematical models based upon theories of how neurons interact and what sort of controls are needed to run a set of legs, simulate them by computer, and see what behaviour they produce. Or we might try to work backwards, from the observed gait patterns, and guess what general kind of circuitry must be involved.

Trying to locate a specific piece of neural circuitry in an animal's body is like searching for a needle in a haystack. Indeed, far worse: the brain contains about 10^{10} neurons and there are at least 10^{12} connections between them, whereas a haystack could contain no more than 10^9 needles, and that only if there were no hay. The physicist Richard Feynman recounts an occasion when, having got interested in biology, he went to the library to ask for 'a map of the cat'. 'A map of the cat sir? You mean a zoological chart!' said the horrified librarian, and for months everyone was talking about the dumb student who wanted a map of the cat. Despite the danger of upsetting librarians, what we'd really like is a *neural* map of the cat: it would be a fantastic breakthrough! In fact, for some very simple creatures, notably the nematode worm *Caenorhabditis elegans* which we encountered in chapter 7, the structure of the entire nervous system is known, cell by cell. There are 959 cells in every such worm, and most of them aren't nerve cells. Even so, tracking all the neural connections is a tough job. Given such a 'circuit diagram' – a 'map of the worm' – you can figure out which bits do what, and how they do it. But to map out the nervous system of a cat is beyond the capability of today's science.

Moreover, unlike nematodes, cats don't all have the same number of cells, and they aren't all built to the same pattern; so while you might work out the circuit diagram of *a* cat, there's no such thing as a circuit diagram for 'the' cat. That doesn't mean that you can't make a sensible guess as to where a CPG might be located, and look there, but despite a widespread belief that CPGs exist, the direct evidence is slight.

Designing specific mathematical models is the traditional approach, and this is where most work has been done. But looking for general principles and model-independent phenomena is more in the spirit of this book. In any case, it's very useful to know what the model-independent features of gaits are. Those provide a general framework for more detailed analysis, and help us avoid the trap of treating features that we don't realize are model-independent as evidence in favour of some specific model.

For example, suppose somebody were to set up a very specific, enormously complicated system of equations, modelling the gait of a biped. Most probably those equations would describe a system with two identical or symmetrically related components, one for each leg, because that's a natural assumption for any modeller to make – probably without even realizing it explicitly. A massive computation based on those equations might well lead to the conclusion that out-of-phase motion can occur. At first sight that might look like strong evidence in favour of the model, because we know that out-of-phase motion *does* occur! However, if you know that virtually *any* model that consists of two identical subsystems coupled together is capable of producing out-of-phase motion, then you realize that out-of-phase motion is evidence only for the general form of the model, not for any specific details. That's not to say that you can't test the details of the model too; but it shows where the emphasis should be placed.

What we're looking for, therefore, is the simplest type of circuit that might produce all the distinct but related symmetry-patterns of gaits. At first sight this looks like a tall order, and we might be forgiven if we tried to concoct some elaborate structure with switches that effected the change from one gait to another – very much a 'gearbox' model!

But there's a simpler and more natural way. The patterns observed in gaits are strongly reminiscent of symmetry-breaking Hopf bifurcations, which in turn suggests that we should think about symmetric networks of neural oscillators. As we've seen, such networks naturally possess an entire repertoire of symmetry-breaking oscillations, and naturally switch between them at bifurcation points. You don't need a complicated 'gearbox'.

Two Oscillators

Just as described in chapter 3 for polygonal networks of oscillators, the mathematics of symmetry-breaking lets us classify the typical space-time symmetries that arise when a CPG network starts to oscillate. For example, let's begin with the simplest possible network: two identical 'cells' (Figure 8.10). These mathematical 'cells' can be

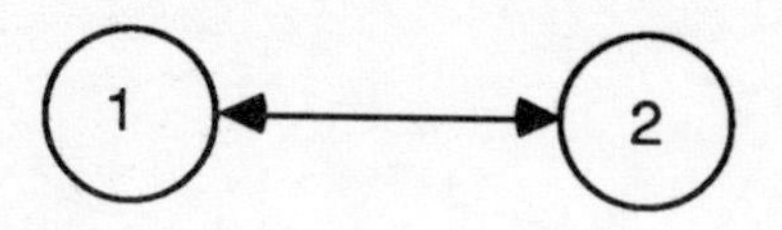

Figure 8.10 Network of two coupled cells (schematic)

entire systems of biological ones, forming complex neural circuits: as far as the symmetry properties go, we're not interested in their internal structure, but only on how they're coupled together. This would most naturally control the motion of a biped, with one cell (neural subcircuit of the CPG) controlling each leg.

If two identical oscillators are coupled together then there are two typical oscillation patterns:

- The *in-phase* pattern: both oscillators have the same waveform. That is, if we interchange oscillators ($1 \rightarrow 2$, $2 \rightarrow 1$) then the resulting dynamics is identical.
- The *out-of-phase* pattern: both oscillators have the same waveform except for a half-period phase difference. That is, if we interchange oscillators ($1 \rightarrow 2$, $2 \rightarrow 1$) then the resulting dynamics is identical except for a phase shift of half a period.

These patterns are illustrated schematically in Figure 8.11. They can arise through a type of Hopf bifurcation. When a system has a reflectional symmetry, then there are two types of Hopf bifurcation – one that gives rise to an in-phase periodic solution and one that gives rise to an out-of-phase periodic solution – and typically they're the only patterns that can occur in this manner. We obtained these two patterns for the component oscillators of a network: a CPG. The phases 'lock' together in some fixed manner – here either in phase or out of phase. The signal output by such a network, either as a result of 'spontaneous' activity or in response to a periodic input signal from elsewhere, exhibits the same pattern of phase-locking.

Suppose we use this output signal to 'drive' the muscles that control a biped's legs, by assigning one set of muscles to each oscillator. The network has these two natural oscillation patterns, but what about the resulting gait? It inherits the two patterns. For the in-phase oscillation of the network, both legs move together: that is, the animal performs a two-legged hopping motion, as in Figure 8.12. The out-of-phase motion of the network produces a gait resembling the human walk: both legs do the same thing, but half a period out of

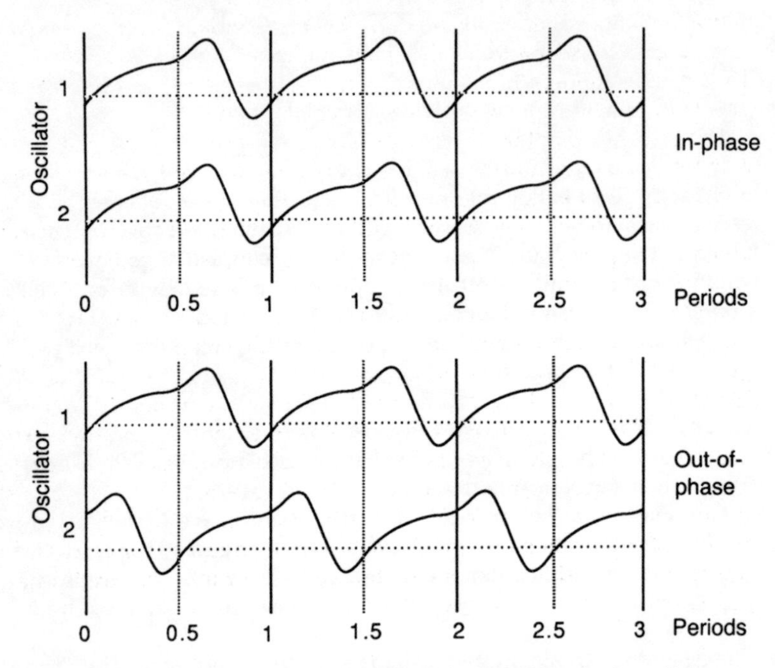

Figure 8.11 Oscillation patterns for two coupled cells. Above: in-phase. Below: out-of-phase

Figure 8.12 In-phase bound of kangaroo

phase with each other. We've already seen that *precisely* these two gaits are observed in bipeds! So here the simplest possible symmetric neural network produces the 'correct' in-phase and out-of-phase gaits as a natural consequence of the mathematics of symmetry-breaking. This result is model-independent; that is, it doesn't depend upon the detailed dynamics of the individual oscillators.

The price we pay for thinking purely about symmetries is that the result gives us no information about the dynamics of the individual oscillators! That's the whole point: *model-independent results don't depend on the model*. But the results do tell us that a network of two identical coupled circuits can do the job, so now we know where to look for more detail. In other words, model-independent results let us work out the general type of model that is needed; at that stage we can plug in model-dependent features to refine our theory.

As well as the human walk, another gait, known as the *half-bound*, also has the out-of-phase symmetry (Figure 8.13). On the first half of the cycle, the left leg hits the ground before the right; but on the next half-bound, the right leg hits before the left. So it takes two half-bounds to complete the periodic cycle. This example shows that symmetries alone may not always distinguish gaits uniquely: there are finer classifications in which gaits with identical symmetries are considered to be different. Taking the duty factor into account is one possibility.

Figure 8.13 Out-of-phase half-bound of Severtsov's jerboa

Four Oscillators

According to Mr Squeers in *Nicholas Nickleby*, the word 'quadruped' is 'Latin for 'orse'. In fact the horse is one of the most versatile quadrupeds, using a remarkable variety of gaits. Does each gait need its own network, or can a single type of network produce all the necessary gaits? The two-cell case leads us to hope that a single network can generate lots of different gaits: this is an appealing theory because it cuts down the complexity of the circuitry required by a CPG. What should the layout of the network be? In some quadruped gaits all four legs do different things, so it looks as if we need at least as many oscillators as there are legs. The simplest model is thus a system of four distinct but coupled oscillators (where by 'oscillator' we really just mean 'subcircuit': it only has to oscillate when coupled up to the rest of the circuit). You can build a lot of circuits with four oscillators, because they can be coupled in a lot of different ways, so we need a systematic approach. Neural circuits can be built to any reasonable pattern, but our choice of networks is guided by the idea that the structure of the CPG should to some extent mimic the gross physiological structure of the whole animal. You don't *have* to make that assumption, but it offers a natural match between the oscillations of the CPG and 'comfortable' motions of the animal's body.

A network of oscillators can be represented by a diagram, in which the oscillators are drawn as various types of blob, and the connections between them (or interactions between them) as various types of line joining the blobs. Different types of blob represent different types of oscillator; more importantly, oscillators represented by the same type of blob are identical in behaviour. Similarly, different types of line represent different types of connection; but blobs connected by the same type of line connect in the same way.

Since we don't really know what the circuit should look like, we'll take a look at five possible cases, represented graphically in Figure 8.14. Here there are either one or two types of oscillator, shown by circles and squares, and up to three distinct types of coupling between them, shown by lines of various kinds. Note that in case 3 all four front-back links are identical, even though some have to be drawn longer than others. The five cases are chosen bearing in mind two alternative interpretations: either as simple networks of cells in a CPG, or as some coarse model of the physiological type of the animal. Indeed one intriguing aspect of our results is that there are natural similarities between the two interpretations, perhaps suggesting a

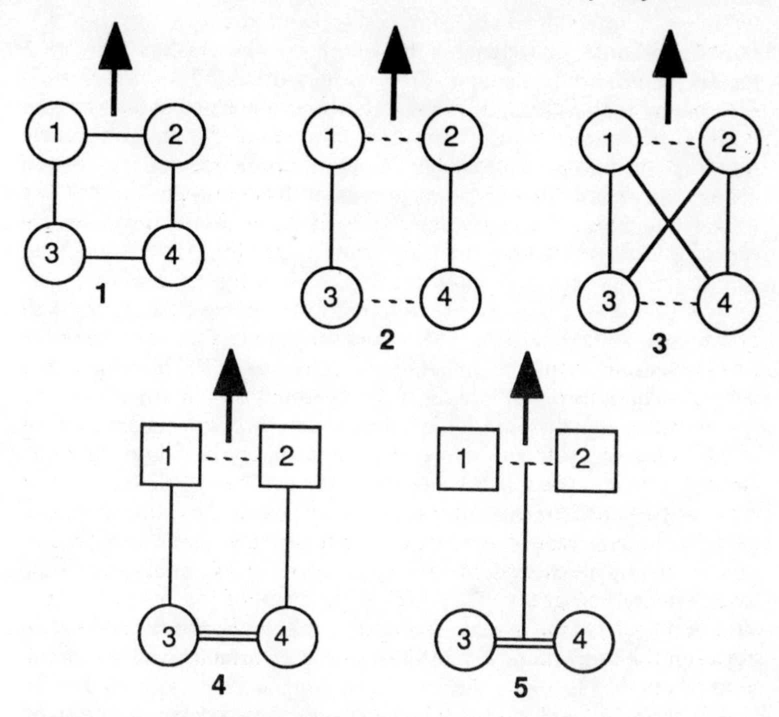

Figure 8.14 Five possible networks of neural oscillators. The animal is facing in the direction of the arrow, and limbs, viewed from above, are controlled by the corresponding oscillator

natural evolutionary route for the development of controlled locomotion.

For networks of oscillators, the symmetries that we've previously referred to as 'spatial' are *permutations* of the oscillators. A permutation is a transformation that rearranges things. Imagine that the blobs are coloured buttons (one colour for each different type of oscillator) and the lines between them are coloured pieces of elastic (one colour for each type of link). Then a symmetry of the network is a way to rearrange the buttons, carrying the elastic with them, so that after the rearrangement the entire network, colours and all, looks the same as it was to start with. (We use mathematicians' elastic which can pass through other bits of elastic, or buttons, without breaking.)

Network 2, for instance, has the same symmetries as a rectangle: we can rotate it by 180°, or reflect it in the horizontal or vertical axes.

Or leave it unchanged, of course. We can't rotate it through 90°, though, because that changes the colours of the elastic. Network 1 has the symmetry of a square. So does network 3, but in disguise. At first it may look as though it has only the symmetries of a rectangle; but we can swap the pair 1 and 2 without changing the form of the network, or the pair 3 and 4. The front and back edges of network 3 are disguised versions of the diagonals of the square.

The five networks correspond naturally to certain physiological types of animal. We take them in turn.

- Network 1 is an arrangement of four identical oscillators with square symmetry. We're not suggesting that there exist animals with square symmetry! Only the CPG is to have that symmetry: four identical cells, coupled in a ring. This arrangement would be appropriate for animals where all four legs are approximately the same and where the coupling between them is relatively similar.
- Type 2 has rectangular symmetry, and suits the physiology of animals whose front and hind legs are fairly similar, but where the left/right coupling differs substantially from the front/rear.
- Type 3 is interesting because it preserves the differences in coupling of type 2, but has more symmetry. The neural circuit effectively treats the two front legs as a unit, coupled to an identical unit at the rear. Its symmetries independently swap oscillators (12) and (34) in pairs, together with the interchange of front and rear. If, for example, front and rear legs are mainly coupled through the spine, then type 3 might well be appropriate.
- Types 4 and 5 are analogous to 2 and 3, but seems most appropriate for creatures in which the front pair of legs differs substantially from the back pair.

For these five networks, just as for the two-cell network, the typical symmetries of periodic oscillations that are always created by Hopf bifurcation can be calculated using general principles of symmetry-breaking. To give you a feeling for what kind of information the theory provides, the results are shown in table 8.1. Here the symbols A and B are used to denote possibly different oscillations (waveforms), and $A + \frac{1}{2}$, for example, means waveform A phase-shifted by half a period. For some particular choices of networks with the given symmetries, there may be additional patterns, but the existence of those listed is model-independent. Each network has its own particu-

Table 8.1 Typical patterns for coupled nonlinear oscillators

System	LF	RF	LH	RH	Comments	Corresponding gait
1a	A	A	A	A		pronk
1b	A	A + $\frac{1}{2}$	A + $\frac{1}{2}$	A		trot
1c	A	A + $\frac{1}{4}$	A + $\frac{3}{4}$	A + $\frac{1}{2}$		similar to rotary gallop
1d	A	A + $\frac{3}{4}$	A + $\frac{1}{4}$	A + $\frac{1}{2}$		similar to rotary gallop (opposite orientation)
1e	A	A	A + $\frac{1}{2}$	A + $\frac{1}{2}$		bound[a]
1f	A	A + $\frac{1}{2}$	A	A + $\frac{1}{2}$		pace
1g	A	B	B + $\frac{1}{2}$	A	A = $\frac{1}{2}$ period	
1h	A	B	B	A + $\frac{1}{2}$	B = $\frac{1}{2}$ period	
2a	A	A	A	A		pronk
2b	A	A + $\frac{1}{2}$	A	A + $\frac{1}{2}$		pace
2c	A	A	A + $\frac{1}{2}$	A + $\frac{1}{2}$		bound[a]
2d	A	A + $\frac{1}{2}$	A + $\frac{1}{2}$	A		trot
3a	A	A	A	A		pronk
3b	A	A	A + $\frac{1}{2}$	A + $\frac{1}{2}$		bound[a]
3c	A	A + $\frac{1}{2}$	A + $\frac{3}{4}$	A + $\frac{1}{4}$		walk and amble
3d	A	A + $\frac{1}{2}$	A + $\frac{1}{4}$	A + $\frac{3}{4}$		
3e	A	A + $\frac{1}{2}$	A + $\frac{1}{2}$	A		trot
3f	A	A + $\frac{1}{2}$	A	A + $\frac{1}{2}$		pace
3g	A	A	B	B + $\frac{1}{2}$	A = $\frac{1}{2}$ period	
3h	A	A + $\frac{1}{2}$	B	B	B = $\frac{1}{2}$ period	
4a	A	A	B	B		asymmetric bound[a]
4b	A	A + $\frac{1}{2}$	B	B + $\frac{1}{2}$		
5a	A	A	B	B		asymmetric bound[a]
5b	A	A + $\frac{1}{2}$	B	B		
5c	A	A	B	B + $\frac{1}{2}$		

[a]Bound is close to transverse and rotary gallops.

lar set of 'natural' oscillation patterns. A few of these possess a remarkable feature, which we explained in chapter 3: some oscillators are required to have half the period of the rest! This property is equivalent to the oscillator being half a period out of phase with itself, which roughly speaking is where it comes from. We'll shortly see that precisely these 'exotic' patterns appear not to occur in normal quadruped gaits. What about abnormal gaits? Wait for the tale of the three-legged dog!

The most symmetric gaits (pronk, trot, bound, pace, walk) correspond very well to patterns in table 8.1. The final gait listed for type 1 has the correct phase relations for a canter, but in other respects is highly unlike that gait. Rotary and transverse gallops aren't on the list at all, although type 1 has two patterns, 1c and 1d, that are similar to the rotary gallop. To find gallops, we may need to look elsewhere. In 1989 Peter Ashwin and Jim Swift found systems of four coupled oscillators in which gallop-like gaits are natural, using oscillators with an extra internal 'reflectional' symmetry. Lots of real-world oscillators have this extra symmetry: the pendulum is an example, and legs aren't too far removed from pendulums. The natural patterns now have diagonally opposite pairs of oscillators half a period out of phase, and there's an *arbitrary* phase difference between the two diagonal sets. Jay Alexander and Giles Auchmuty discovered the same result in a more special model in 1983.

The canter is something of a puzzle, although it may have an explanation along similar lines to the two gallops. There's something strange about the canter: most horses have to be *trained* to do it – the gait doesn't come especially naturally. A possible mathematical explanation might be based upon a CPG with tetrahedral symmetry: teaching a horse to canter may amount to teaching it to operate all four legs independently.

Leaving the canter aside, we've shown that all of the gaits listed earlier have the same symmetries as natural oscillations of simple, symmetric neural networks. Moreover, a large proportion of the natural oscillations of the networks studied correspond to observed gaits. Indeed among the twenty patterns listed for the first three types (all oscillators identical), only three don't seem to correspond to observed gaits. All of these involve the curious 'half-period' condition.

Each individual animal has its own set of characteristic gaits, its own physiology, and its own CPG. No particular network has to produce *all* possible gait patterns: indeed it would be embarrassing if one did! All it has to do is produce the range of gaits observed in the corresponding animal. For example, animals that (say) walk, bound, trot, and pace (and – rarely – pronk) might possess a CPG of type 3, and so on. Moreover, useful gaits must possess other properties than just being natural oscillations: they should use energy efficiently, not make the animal fall over, and so on. Extra considerations like these presumably rule out those oscillations patterns that aren't observed in practice – though it would be nice to understand exactly why and how.

Does our guess that the CPG structure is to some extent related to the animal's physiology hold up? The gaits observed in animals like the horse, whose front and back legs are very similar, occur in networks 1, 2, and 3, which we have already said are physiologically appropriate in such cases. In contrast, the patterns for types 4 and 5 are all plausible for creatures whose front legs are very different from their rear legs. For example 5b corresponds to a two-legged walk on hind legs, while the front legs move together; 4b is essentially the normal human walking gait with B representing arm-movements and A leg-movements. So there's quite a good correspondence between physiology and CPG network.

The Three-legged Dog

As well as matching up observed gaits with CPG network models, we can make predictions about more unusual systems. For example, Hopf bifurcation with triangular symmetry tells us what gaits to expect in a three-legged animal: the four patterns listed in chapter 3.

But there aren't any three-legged animals.

We thought so too – until recently. Then Peter Ashwin reported seeing a three-legged dog being exercised in a park in the town of Kenilworth. It had lost one hind leg in an accident. It compensated for this by angling the leg towards the centreline; and it adopted two gaits. At high speed, the front legs moved in phase, and the rear was pretty much out of phase, a sort of bound. This is consistent with the oscillation pattern for a triangular network that we called the Reflectionally Symmetric Oscillation in chapter 3. The most striking thing was the second gait, which it used at lower speeds. The two front legs were half a period out of phase, like a human walk; and the back leg hopped up and down *twice as rapidly*. That is, the back leg moved with half the period of the front legs. This is another typical pattern for a triangular network – we called it the Twisted Reflectionally Symmetric Oscillation – and it has the strange half-period property.

This is a very curious vindication of a very curious mathematical prediction.

As it happens, there's a simple way to rationalize the double frequency movement of the three-legged dog's hind leg. For balance he must move his hind leg every time he moves one of his forelegs. Indeed, the same pattern can be observed in a person who must use a cane to walk. Often they must move the cane (or third leg) each time they move one of their legs.

All Change

So much for the symmetries of individual gaits. We now tackle another important question: why do gaits *change*, and why do their symmetries change too? Symmetry-breaking provides a natural reason for expecting such changes. As we've already seen, a common phenomenon in dynamical systems is bifurcation. A bifurcation is a qualitative change in the dynamics, occurring when parameters are varied. For our present purposes the most important type of bifurcation is Hopf bifurcation, described in chapter 3, for which a steady state changes to a periodic state. Hopf bifurcation is a very common way for a periodic oscillation to 'switch on'. When several different oscillatory motions are involved, several different Hopf bifurcations may be involved, and the full set of bifurcations can become very complicated. Figure 8.15 is a schematic representation of a typical sequence of bifurcations in a four-oscillator network, labelled with the corresponding gaits. The curves represent possible states of the system and the branch-points are transitions between them. The curves that branch from the basic horizontal line (representing no motion) are called *primary modes*; the others are *mixed modes* and represent a kind of superposition of the two primary modes that they join. The horizontal coordinate is some measure of the speed at which the animal 'wants to move': a variable controlled by the brain and affecting the behaviour of the CPG network. We can't employ the *actual* speed in this role since this is a consequence of the dynamics,

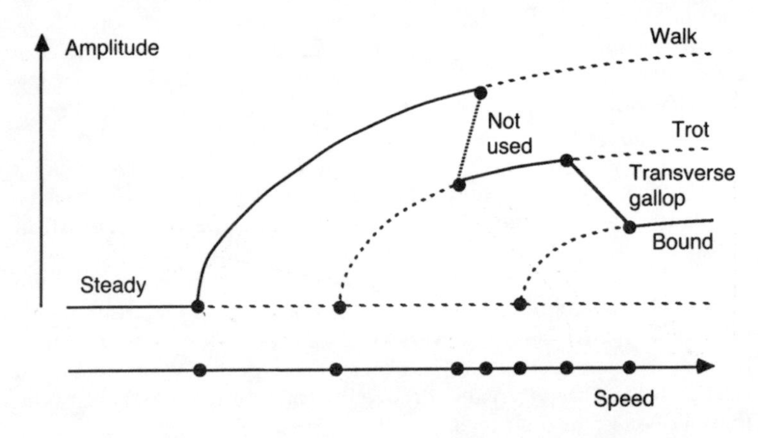

Figure 8.15 Schematic picture of sample bifurcations in animal gaits

not a cause of it. From now on, whenever we say 'speed', it's with this type of intepretation in mind.

The sequence of events described by this picture, for increasing speed, corresponds to the upper curve in Figure 8.16. It is:

steady → walk ⇒ trot → transverse gallop → bound

where → indicates a continuous transition and ⇒ a jump. An interesting feature emerges if we reverse the process, so that the animal slows down again. Now we have the lower curve in Figure 8.16, which corresponds to the sequence:

bound → transverse gallop → trot ⇒ walk → steady

in reverse order. However, the speeds at which some transitions occur can be different when the speed is reduced rather than increased. This effect, known as *hysteresis*, is associated with jump transitions and doesn't occur for continuous transitions. Hysteresis occurs in a car with a reluctant driver who always changes gear too late. Suppose that a change from second to third gear should ideally occur at 30 mph. The driver is late changing up a gear, so changes up from gear 2 to 3 at 35 mph, but is also late changing down, so the change from gear 3 to 2 occurs at 25 mph. The two speeds at which the gear change occurs are different, depending on whether the speed of the car is increasing or decreasing.

In general, the possible transitions are organized by their symmetries. The faster the animal moves, the less symmetry its gait has: more speed breaks more symmetry. This is consistent with the general phenomenon that the more a system is 'stressed', the less

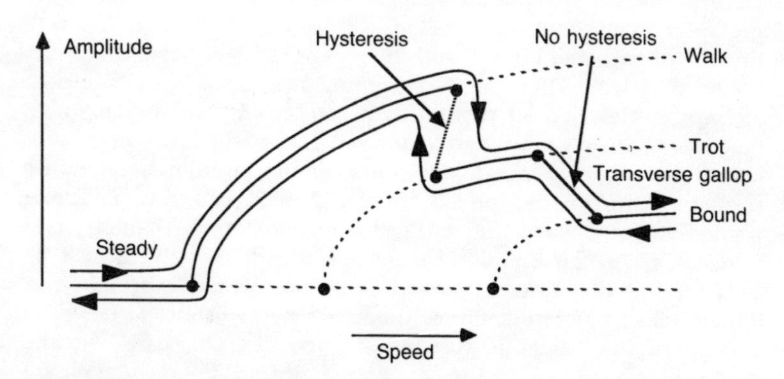

Figure 8.16 Hysteresis in animal gaits

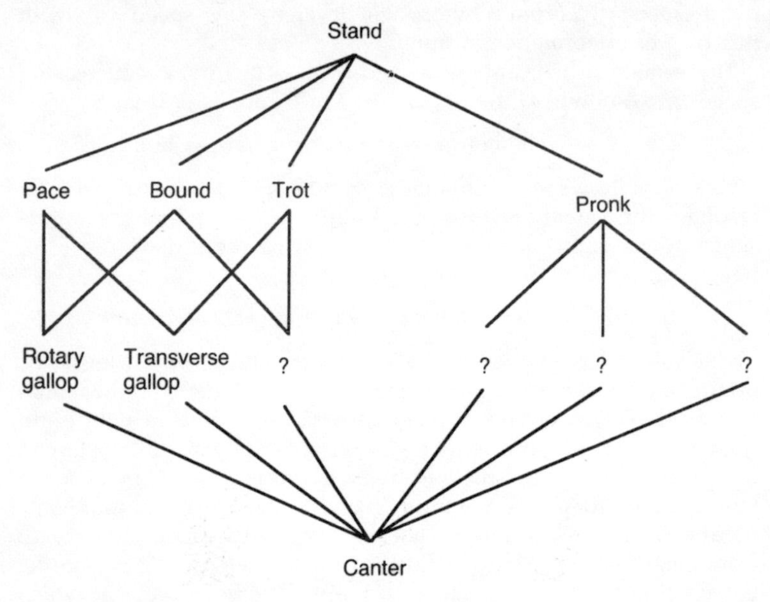

Figure 8.17 A universal pattern of symmetry-breaking in animal gaits. Arrows indicate loss of symmetry

symmetry its states possess. Using the model of rectangular symmetry, type 2, the theoretical patterns of symmetry-breaking are shown in Figure 8.17. We see that the primary modes are the most symmetric gaits (pronk, pace, bound, trot) and the secondary 'mixed' modes are the rotary and transverse gallops. Indeed the rotary gallop is a mixture of pace and bound, and the transverse gallop is a mixture of bound and trot. There is an intriguing resonance with Fabricius' attempt to describe all motion as a combination of walking with diagonal limbs in unison and leaping with opposite limbs in unison.

Is there any evidence that gait transitions in animals really are related to bifurcations? Indeed. In 1981 D. F. Hoyt and R. C. Taylor trained horses to walk, trot, and gallop on a treadmill. Their observations of oxygen consumption against speed are shown in Figure 8.18. This type of diagram is precisely what one would expect from a dynamical system bifurcating into one of several distinct modes. The overlap of the parabolic curves for trot and gallop suggests the presence of hysteresis in the corresponding transition. The data are more equivocal for the walk/trot transition but support a smaller

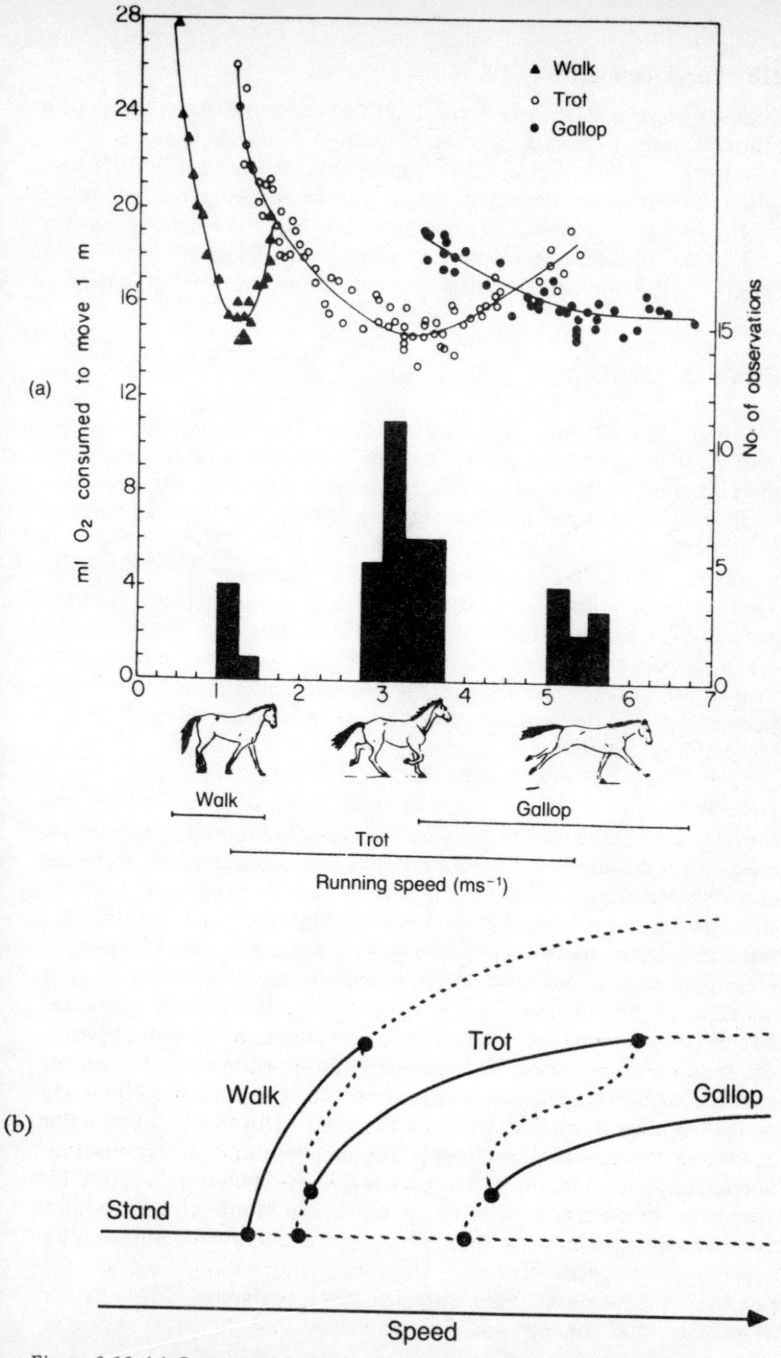

Figure 8.18 (a) Oxygen consumption (vertical axis) of horses for various gaits and speeds (horizontal axis). (b) Interpretation as a bifurcation diagram, same horizontal scale

degree of hysteresis: the interpretation of these results in terms of a bifurcation diagram is shown at the bottom of the picture.

A further interesting discovery made by Hoyt and Taylor is that when the horses are permitted to select their own speeds, depending on the terrain, they tend to minimize their oxygen consumption for each gait. They therefore select specific speeds from the continuous range that's available. As Fabricius said, nature does what is best.

Centipedes Revisited

Whatever the physiological constraints on locomotion, the remarkable variety of natural gaits basically boils down to the standard mathematical patterns of symmetry-breaking in networks of coupled oscillators. This is a valuable unification; moreover the idea extends to other types of movement and to other creatures. The flight of birds involves bilaterally symmetric oscillations of two appendages – wings. Unlike the passenger jet of chapter 1, the natural pattern of flight doesn't break the bird's bilateral symmetry (Figure 8.19). Of course, the aerodynamics of flight is a far more subtle problem than its symmetry. On the ground, some birds hop – again preserving their symmetry – but others walk, breaking it to the out-of-phase pattern.

The theory extends readily to hexapods – six-legged creatures, such as insects. For example, the typical gait of a cockroach, and indeed of most insects, is the *tripod*, in which the middle leg on one side moves in phase with the front and back on the other, and then the other three legs move together, half a period out of phase with the first set. Mathematical analysis (Figure 8.20) reveals this as one of the natural patterns for six oscillators arranged in a hexagon. Myriapods (centipedes and millipedes) produce rippling patterns of leg-movements (Figure 8.21). These can be understood as travelling waves in large networks with polygonal symmetry, corresponding to (mathematically!) gluing the creatures' front ends to their rears to keep the wave travelling. The movement of fish, lizards, worms, and snakes can be described in similar ways (Figure 8.22). Even some types of protozoon – microscopic single-celled creatures – propel themselves along by rotating a helical tail, or flagellum, just like the mechanical device known as an Archimedean screw. This movement has rotating wave symmetry akin to that of spiral flow in the Couette–Taylor system.

Perhaps Mrs Craster's centipede just needed to bone up on symmetry-breaking.

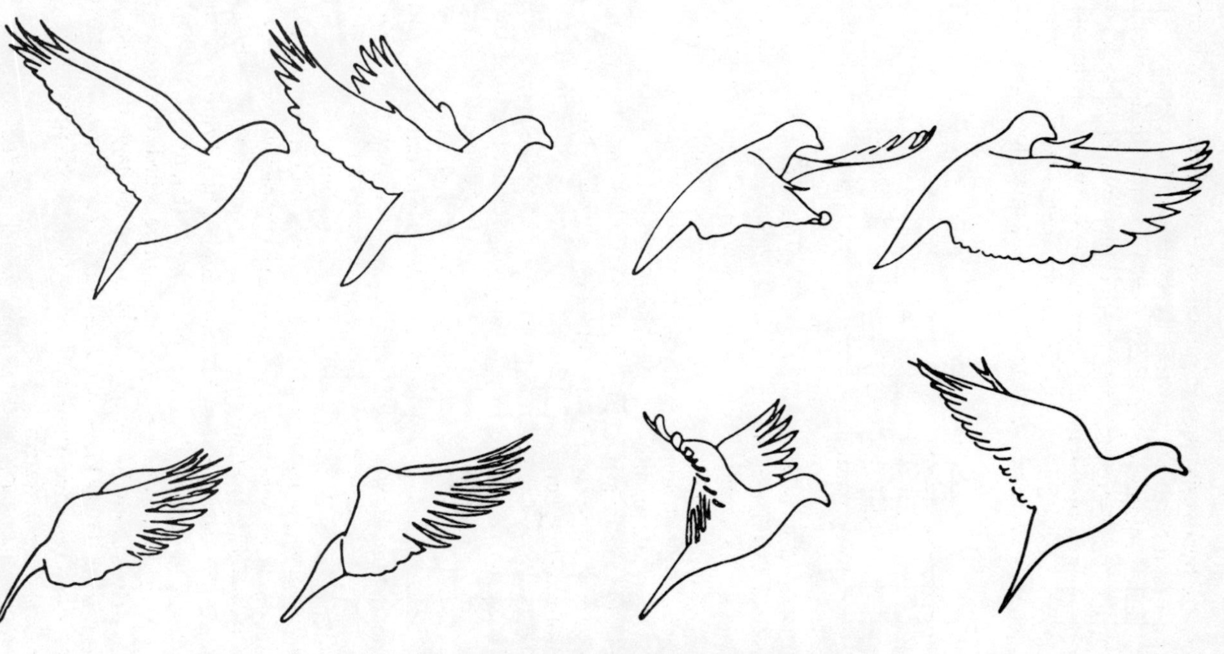

Figure 8.19 The in-phase flight of a pigeon

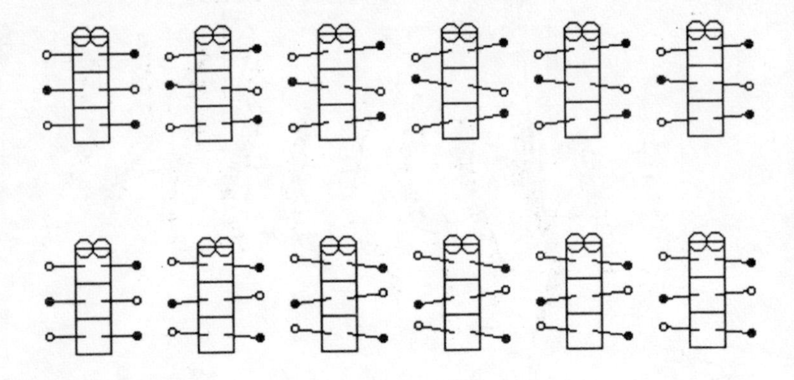

Figure 8.20 Simulated tripod gait of a computer bug. White legs move in phase; black legs are half a period out of phase with white legs

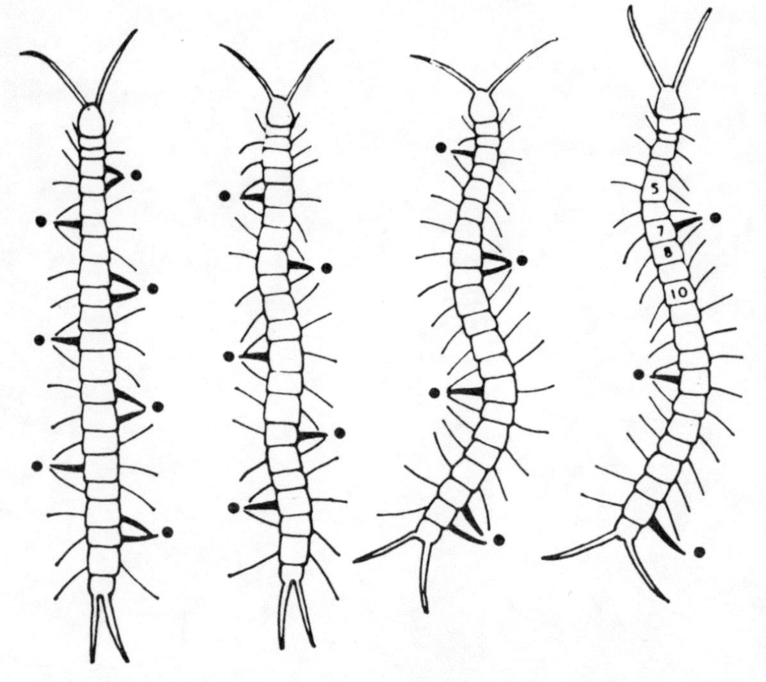

Figure 8.21 Centipede gaits in the wild. Speed increases from left to right. Thick lines indicate legs in contact with the ground

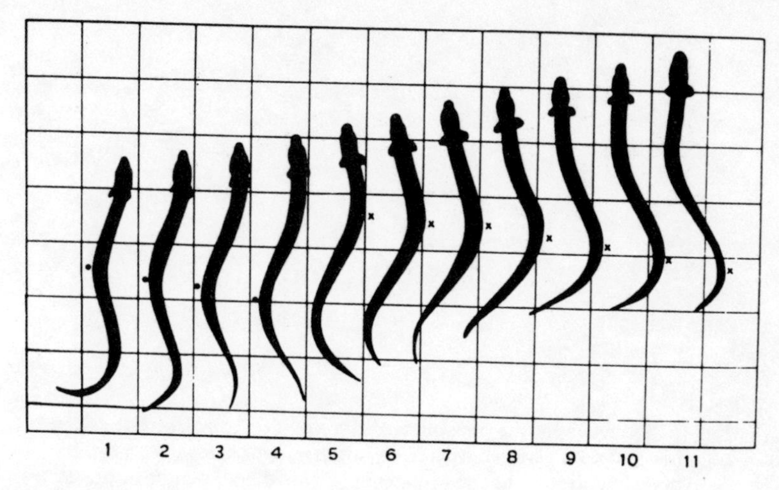

Figure 8.22 Travelling wave in a moving butterfish

9

Icons of Chaos

> But in honeycombs the reason is different. For bees are not rolled together pell-mell, like cells in a fruit, but marshall themselves as in line of battle, all with their heads projecting in one direction or its opposite, as they think best, all pushing in mutual support tail to tail.
>
> Johannes Kepler, *The Six-Cornered. Snowflake*

It's time now to pick up a thread we left hanging in our discussion of the Couette–Taylor experiment in chapter 5: turbulent Taylor vortices. (Take a look at Figure 5.2d) to remind yourself what it looks like.) What a surprise this fluid state is! A fluid that is being stressed hard enough so that its motion is turbulent – yet still the fluid is able to exhibit a most distinctive pattern, the same pattern that appears in that much simpler motion, Taylor vortices. We have a signature of order in chaos.

Our emphasis, hitherto, has been on symmetry-*breaking*. Most of the patterns that we've described have formed through a breakdown of symmetry. Taylor vortices bifurcate from the laminar, unstructured Couette flow, losing symmetry but gaining structure. In a different way, time-period spiral vortices form from time-independent Couette flow by a breakdown of both spatial and temporal symmetry. A cell divides, loses symmetry, and gains structure. Animal gaits are discernible patterns of motion that can be understood as a breakdown of symmetry.

This kind of pattern-formation has its *own* pattern: structure is gained as symmetry is lost. Indeed the general picture that's emerged so far is that, as a system is put under greater stress, the observed states lose symmetry and gain in complexity (both spatial and temporal). But what about turbulent Taylor vortices? As the Couette–Taylor system is put under greater stress, the patterned but time-independent Taylor vortices (Figure 5.2b) bifurcate from the homogeneous Couette flow (Figure 5.2c); then time-periodic wavy vortices (Figure 5.2c) bifurcate from Taylor vortices; then modulated wary vortices with two independent frequencies bifurcate from wavy vortices; and eventually there's a complete breakdown of temporal symmetry to a turbulent state yielding very complicated motion. But – the Couette-Taylor experiment isn't finished! As the system is further stressed, a state with more symmetry *and* more structure (Figure 5.2d) appears. This seems quite contrary to everything we think we've understood.

Let's examine the sequence of events more closely. The symmetry of turbulent Taylor vortices isn't exact. If you look at the fluid state closely, then you'll see imperfections in the pattern that make it clearly distinguishable from Taylor vortices. On average, however, the pattern is symmetric. What we find is a state that in some sense seems to exhibit more symmetry – at least on average – than did the state from which it emerged. At present there's no proven explanation of what's happening in this form of symmetry-creation. We can't even throw it on a supercomputer and tell it to analyse the inner workings of turbulent Taylor vortices, because the computation of turbulent vortices from the Navier-Stokes equations is well beyond the limits of current supercomputer power. So we're on our own: human imagination versus the infinite.

When faced with a complicated issue it's often best to ask (and to answer) the simplest possible questions – just to get a sense of where you might be heading. There's a well-developed mathematical theory of complicated dynamics, the theory of *chaos*. It forms the subject matter of the predecessor of this book, *Does God Play Dice*? There's no question that turbulent motion is complicated – at least as complicated as chaotic dynamics. Indeed the prevailing theory is that turbulence is a physical manifestation of chaotic dynamics. Given that it's impossible to study the effect of symmetries on turbulence directly, it seems reasonable to ask what effect symmetries have on chaotic dynamics.

Can symmetry, or pattern, coexist with chaos, or disorder?

The two requirements seem contradictory. Yet the answer is 'yes', and it leads to a topic that's well worth exploring in its own right.

Fruit Flies on the Farm

Or, as the saying goes, fruit flies like a banana. In fact, you're now going to get a quick introduction to the topic of chaos, without symmetries. For more details, see chapter 8 of *Does God Play Dice?*

Suppose we want to model how the population of a given species – say *Drosophila* fruit flies on the Jones farm in Santa Cruz county – varies from one year to the next. The simplest kind of model assumes that the population changes according to a fixed rule. Next year's population is determined precisely by this year's.

But what rule should we choose? An attractive choice is an equation so simple that it seems inconceivable that in 1975 the world would be surprised by its complexity. It's called the *logistic equation*, and while we've tried our best to avoid formulas, for once we feel it makes better sense to write the formula down: it's

$$f(x) = kx(1 - x).$$

First we must explain how this relates to populations. The logistic model assumes that there's a maximum sustainable population on the Jones farm, and uses the value of x to represent the ratio of the actual population to that maximum population. For example, if the maximum population is 1,000 and the actual population is 743 fruit flies, then $x = 0.743$. The number x must therefore always lie between 0 and 1. The number k is a relative growth rate for the fly population, and is assumed not to vary over time. It is called the 'carrying capacity'. The expression $f(x)$ is an estimate for the population of fruit flies on the Jones farm next year, in terms of the population this year. Here it's based on the idea that when the population is low, and there's no competition for space or food, each generation grows relative to the previous one by a factor of k. If this were always the case, we'd just have $f(x) = kx$, or *exponential growth*. But when the population becomes larger, competition cuts down the rate of growth by an amount kx^2, and this becomes more and more important the larger the population x becomes. Putting both effects together we get the logistic equation, which therefore describes exponential growth subject to a cut-off at a fixed population level. The expression f itself is called a *mapping*, because given any point x it 'maps it to' the value $f(x)$.

To use the model we choose an initial population x_1, and then compute next year's projected population $x_2 = f(x_1)$. Then we compute the third year's population $x_3 = f(x_2)$ – and so on. For example suppose $k = 3$, and $x_1 = 0.743$. Then

$$x_2 = f(0.743) = 3(0.743)(1 - 0.743) = 3(0.743)(0.257) = 0.573,$$

and so on. In principle, we can compute an infinite sequence of population predictions: the entire process is called *iteration*. Throughout this chapter we will talk of 'iterating the equation', 'applying the equation to a point', and so on, although strictly speaking it is the mapping f that is iterated or applied, not the equation that it defines. We adopt this unorthodox terminology because equations are familiar to most people, but mappings aren't. The way the numbers x_j change is called the *dynamics* of the population.

The simplest sort of dynamics occurs when the population in year 2 is the same as that in year 1. Because the rule for the next year's population doesn't change from year to year, this means that the population in year 3 is the same as in year 2, and that in year 4 is the same as in year 3 ... and so on. That is, the population remains fixed from year to year. We call this a *steady state*.

The next simplest dynamics occurs when the population in year 2 is different from that in year 1, but that in year 3 returns to the original value that it had in year 1. In such a population the numbers repeat every second year; first the population may be high, then low, and then high again. Such a population dynamic is called a *period two* point. Other periods, such as 3, 4, 5, and so on, are also possible. Indeed, such repetition is not so surprising; plagues of locusts often repeat with great regularity in seven-year cycles. Might this indicate the presence of a period 7 point?

In a paper published in *Nature* in 1976, the mathematical ecologist Robert May noticed that the simplicity of the logistic equation is deceptive. It exhibits all kinds of complicated dynamic behaviour for different values of the carrying capacity k. This behaviour includes period 2 points, period 3 points, period 4 points, and so on – but also points that aren't periodic at all. These *aperiodic points* seem to wander all over the place – one year the population will be low, the next five years high, followed by a year of average population, then a few low years followed by a very high year, and so on – producing any sequence of highs and lows that you care to choose, including a random one. This is *deterministic chaos*. According to the logistic model, the population of fruit flies on the Jones farm can be high some years, low others, average others, and no one can predict which is going to occur.

This may seem like reality! What surprised May was not that real populations might be unpredictable, but that a model as simple as the logistic equation can also produce that kind of complexity. In Figure 9.1 we show the kinds of population dynamics that are possible in the logistic equation. This *bifurcation diagram* plots values of k horizontally, and values of x vertically. To construct the diagram we choose a value of k, run the dynamics for a few steps until it settles

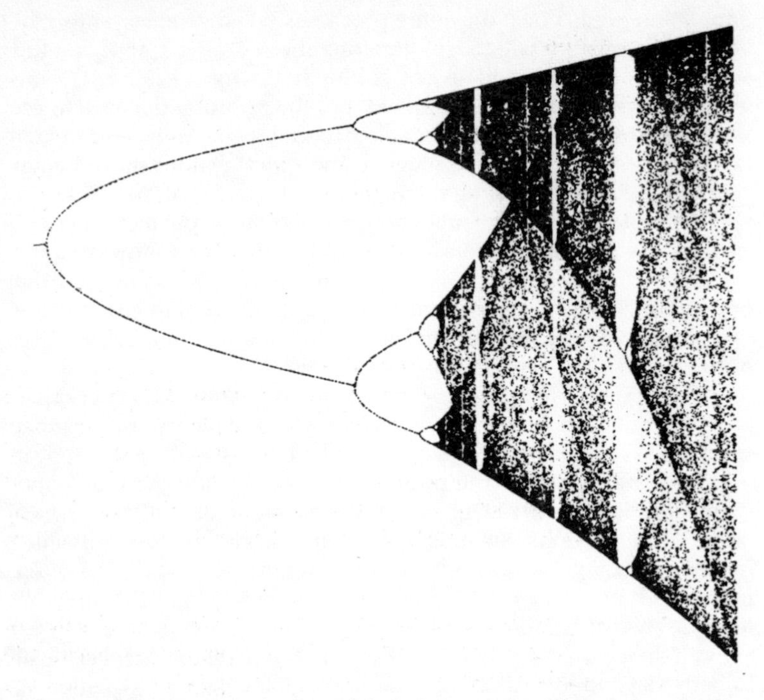

Figure 9.1 The bifurcation diagram of the logistic equation. The carrying capacity k *is plotted horizontally; the long-term behaviour of the population* x *vertically. Curves indicate steady or periodic states, bands are chaos*

down, and then plot a series of x-values vertically above that value of k. We repeat for the next value of k, and so on. Depending on the carrying capacity k we can see regions of fixed points and period-2 points, along with regions of very complicated, chaotic dynamics. In fact, reading the diagram from left to right, we first see a single curve, corresponding to a range of values of k for which there is a single steady state. Then the curve splits (or *bifurcates*, hence the name of the diagram) into two, and for that range of values of k there is a period-2 point. It splits again, corresponding to period-4 points, and so on. The wide bands occur above regions of values for k that produce chaos.

An Odd Equation

We now return to our original question: how can symmetry and chaos coexist? To answer it, we'll perform a numerical experiment in the simplest case we can think of. The simplest symmetry on the line is the reflectional symmetry that takes x to $-x$. An equation having such symmetry is called *odd*; it must satisfy the condition $f(-x) = -f(x)$. The logistic equation isn't odd; but the *odd logistic equation*

$$f(x) = kx(1 - x^2)$$

is. With a name like that, it would have to be.

In this equation the direct analogy with population theory is lost; in particular, when we iterate the odd logistic equation points x are allowed to be negative. Since we're now performing a mathematical experiment, rather than modelling an ecology, we don't much mind that. Figure 9.2a presents the results of iterating the odd logistic equation, just as for the logistic equation. There are several features worth noting.

1 We begin with a positive initial value x_1 and a small value of the carrying capacity k. After plotting points parallel to the vertical axis with that value of k, we change k a little, and then continue to iterate the odd logistic equation starting with the most recent value of x (obtained using the previous value of k).

2 The iterates obtained in this process all remain positive until we reach a particular critical value of k. As it turns out that value is $k_c = \frac{3}{2}\sqrt{3} = 2.598$.

3 Repeating the process described in (1) and (2) but beginning with a negative value of x_1 leads to the diagram shown in Figure 9.2b. The iterates stay negative until that same critical value k_c, where now the iterates go positive.

The set of iterates is always asymmetric when k is smaller than k_c. Whether the set consists of positive or of negative numbers depends only on the choice of the initial value x_1. Above the critical value k_c the set of iterates is symmetric about the origin. The amount of symmetry doesn't decrease, as it does in conventional symmetry-breaking: it increases! What we've witnessed is the simplest example of *symmetry creation*.

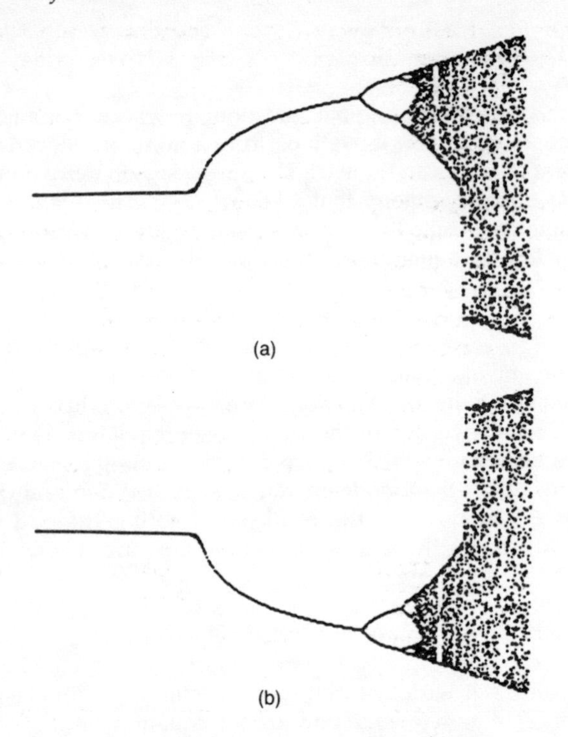

Figure 9.2 (a) Bifurcation diagram for the odd logistic equation, starting from positive initial values of x. *(b) Bifurcation diagram for the same equation, starting from negative initial values*

Note that this symmetry exists only *on average*. If you plot a small number of iterates – say ten – you'll find that they're *not* symmetrically placed about the origin. You can't see the exact symmetry unless you've plotted enough data points. The spherical symmetry of the blastula in a developing cell also exists only on average, in roughly the same sense, and so does that of tree bark; but for chaotic dynamics there's a precise mathematical definition.

Icons and Oscillators

With this example in mind, we can come up with more interesting symmetry groups and ask whether symmetry-creation occurs for these maps too. For example, we can write down equations that define mappings from the plane to itself and have triangular or square or pentagonal symmetry.

Wait a minute! Shapes can have symmetry, but *equations*? Well, in chapter 2 we said that any mathematical structure whatsoever can have symmetry: you just find those transformations that leave its essential structure unchanged. We mentioned Galois theory, where the solubility of equations is determined by their symmetries. So equations *can* have symmetries. However, for dynamics, the essential features of equations aren't the ones that interested Galois, so the appropriate concept of symmetry is different from his. The important feature of dynamics is how a given point is mapped to the next as time clicks one unit on.

Let's think about square symmetry. A square in the plane has eight symmetries; eight rigid motions of the entire plane that leave the square invariant. A 'typical' point in the plane is mapped to eight different images by these transformations (Figure 9.3), forming a

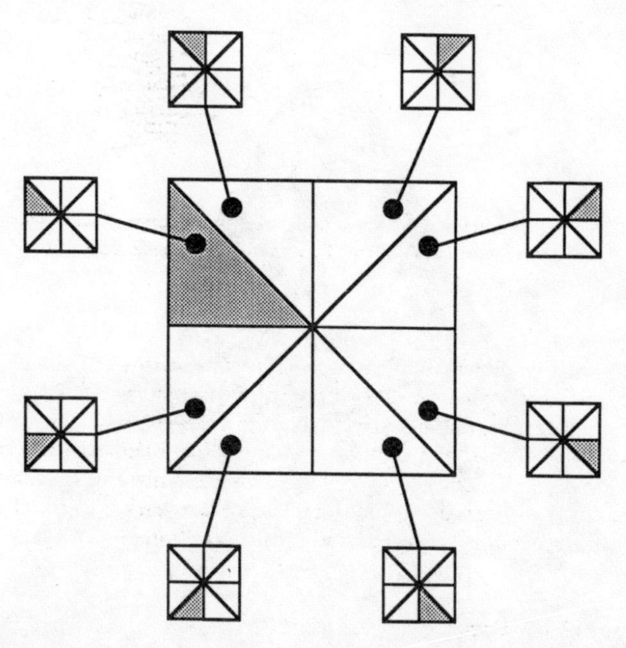

Figure 9.3 The eight images of a point under the symmetries of a square. The original point is in the shaded triangle: small squares indicate the transformations used

shape like a square with its corners lopped off. An equation is said to have square symmetry if, when the equation is applied to each of these eight points – more formally, if the mapping *f* that defines the equation is so applied – it produces another set of eight points of the same type. Here truncated squares must map to truncated squares. Another way to say this is that if you apply the equation to a point and then apply a symmetry, the result is the same as first applying the symmetry and then applying the equation. In short, symmetrically related points have symmetrically related images.

This condition can be used to determine the mathematical form of the equation; and once we've found a symmetric equation, we can use it to do dynamics. We just iterate it in the same way that we did for the logistic and odd logistic equations. First choose an initial point, then apply the equation to obtain a new point, then apply it again to get a third, and so on. To visualize the result, plot all the points on a computer screen. The equations that arise are described in appendix 1. In Figure 9.4 we've reproduced some of results that show

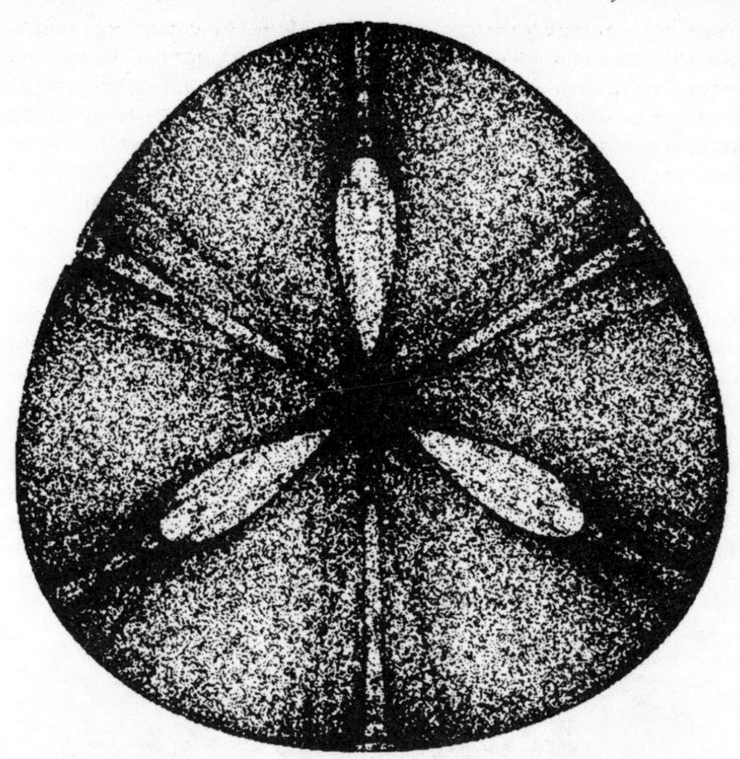

Figure 9.4 Two examples of symmetric icons. See Appendix 1 for details

that symmetry-creation is indeed possible for these more interesting symmetry groups. We call these shapes *icons*.

Having played this mathematical game with equations, we must now ask whether symmetry-creation can be seen in the *real* world. In 1989 Peter Ashwin set out to demonstrate symmetry-creation in the Nonlinear Systems Laboratory at the University of Warwick, under the direction of Greg King. We've already mentioned that systems of coupled oscillators possess natural symmetries. It's also known that oscillators can become chaotic. So Ashwin constructed – as nearly as he could – a system of three identical, identically coupled, electrical circuits (called *Van der Pol oscillators*) and sent an electrical charge through the circuit. Standard AC current is modulated with a frequency of 50 Hz in the UK, and it makes sense to view the result every period of the modulation, which Ashwin did on an oscilloscope. In this way he created an electronic mapping with triangular symmetry that depends on the voltage of the input. Some of the output from Ashwin's circuits is shown in Figure 9.5. These results confirm that symmetry-creation can be observed experimentally. The final, extra chapter of the American paperback edition of *Does God Play Dice?* also describes this work. Big oak trees from little icons grow.

We started with a problem in Couette–Taylor flow: the strange mixture of chaos and pattern that occurs in turbulent Taylor vortices. Much technical mathematical research is still needed before anyone can demonstrate a rigorous connection between this – or any – kind of symmetry-creation, and the type of symmetry-on-average observed in turbulent Taylor vortices. However, we can tell a mathematical story about turbulent Taylor vortices that corresponds precisely to symmetry-creation in chaotic icons. The 'icons' for fluid flow are called *attractors*, and they live in an imaginary mathematical space whose points correspond to all possible flow-patterns. A pattern that changes in time corresponds to a moving point in this space, so it traces out some set. The symmetries of the Couette–Taylor apparatus

Figure 9.5 Symmetric chaos in a system of three coupled oscillators. This picture is processed by computer from experimental data

are inherited by the mathematical space, so we can ask what the symmetry of the attractor that corresponds to a given flow pattern is. For steady states, whose attractors are single points, it's the same as the symmetry of the pattern in its usual sense. But now turbulent states may have a well-defined symmetry. The conjecture is that turbulent Taylor vortices correspond to an attractor that is symmetric under top-bottom reflection and under rotation of the apparatus about its axis, and that it is formed from a fluid state having less symmetry. Should this be correct, it may well explain the symmetry on average that is observed in turbulent Taylor vortices.

In other words, even the turbulent flows possess a kind of symmetry, but to make sense of it we must 'filter out' the fine structure and look only at how the 'turbulent texture' is distributed. If we do this, featureless turbulence has the same symmetry as Couette flow, and turbulent Taylor vortices have the same symmetry as ordinary Taylor vortices. Now the complete 'main sequence' begins by breaking symmetry in several stages, but then (neglecting turbulent fine structure) it restores it all again in reverse order. This sequence of events has a natural interpretation in terms of the associated attractors: the first few stages involve symmetry-breaking, and then the symmetry is put back together again by a series of symmetry-increasing collisions of chaotic attractors. However, to *prove* that this really happens is desperately difficult, and it's not even clear how to get started. Moreover, as we've said, supercomputers don't help.

Whether the goal of finding a complete proof is successful or not, we now have a new phenomenon that has been isolated mathematically, and whose motivation is based on this most interesting fluid dynamic state. But rather than pursuing the technicalities, we now turn our attention to an apparently different kind of intellectual activity – the manufacture of quilt patterns.

Chaotic Quilts

Just think of the many places where you've seen repeating designs. The art of Maurits Escher (Figure 9.6), the wallpaper in your friends' houses, the tile floors of government buildings, Indonesian batiks, Greek mosaics, Islamic designs (Figure 9.7), Scottish tartans and perhaps even the quilt on your bed. There's little question that all cultures have been fascinated by repeating designs.

Indeed, repeating designs are also practical. To display a large picture, you must have the right sized area in which to make that

Figure 9.6 Repeating designs are characteristic of the work of Maurits Escher

display. However, any area, no matter how irregular, may be covered by a small design repeated often enough to fill it. Moreover, such repeating patterns are aesthetically more satisfying when they fit together nicely at the boundaries.

As we've discussed previously, nature too seems to like repeating designs. Recall the structure of regular crystals, the hexagonal shape of honeycombs, and the roll patterns in Rayleigh–Bénard convection.

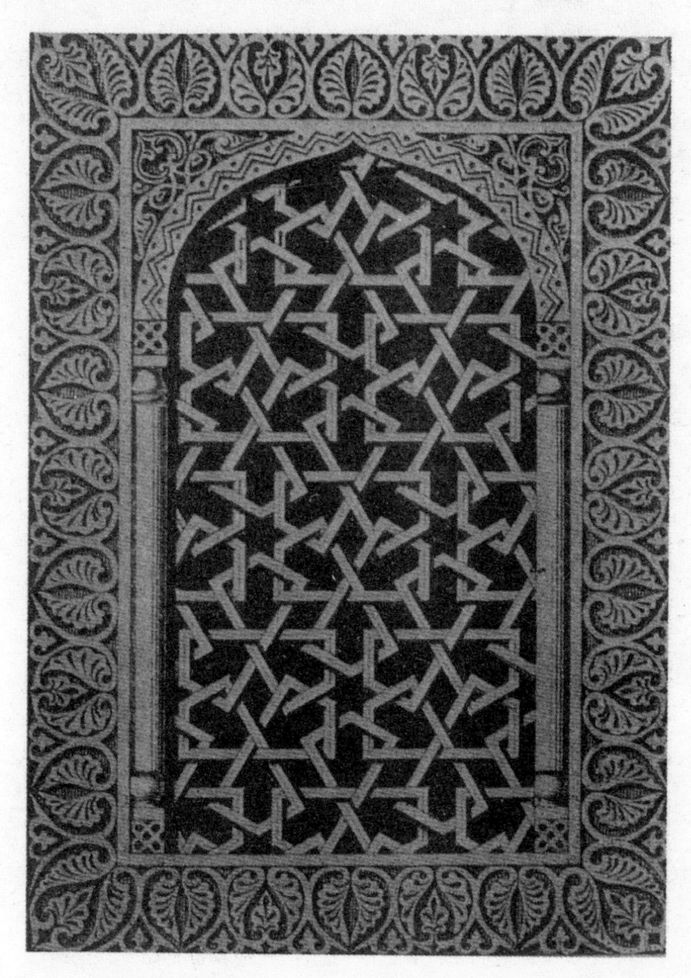

Figure 9.7 An Islamic pattern

With these examples in mind, it comes as no surprise that mathematicians have attempted to classify these repeating patterns. Indeed, the patterns are classified on three levels – type of lattice, symmetry within the lattice, and colour. We've discussed the lattice features already in chapter 4, but not colour. Let's say a bit more.

By a repeating pattern we mean a planar design that can be translated onto itself in two independent directions. Such translations form a *lattice*; for example, there are square lattices, rhombic lattices and hexagonal lattices. We can think of each lattice as comprised of a fundamental *cell* along with two fundamental *translations* that let us tile the plane with translates of that cell (Figure 9.8).

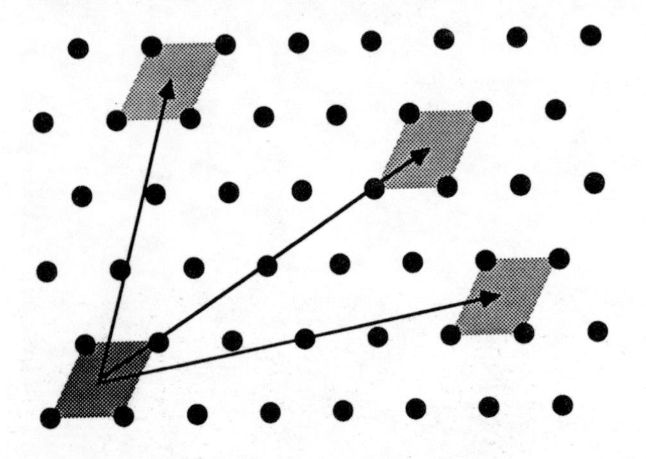

Figure 9.8 Fundamental cell and translations in a lattice

Second, there are the symmetries within each cell. For example, there are patterns that repeat on a square lattice with the design itself not being square symmetric, as well as patterns where the design *is* square symmetric. We can imagine an even more sophisticated pattern by choosing a design that is not rotationally symmetric, and forming a repeating pattern by rotating the fundamental cell clockwise by 90° each time we apply a fundamental translation. See figure 9.9.

Finally, suppose we assume that the design is made using two colours – black and white. Then it is easy to make up a new pattern where black and white are interchanged each time we apply a fundamental translation. Many of Escher's drawings are based on this notion of colour swapping (Figure 9.10). Colour symmetry with several colours is also possible: it was first studied extensively by the Russians A. V. Shubnikov and V. A. Koptsik.

So, after a little thought, it appears that even determining the symmetries of a repeating pattern can be quite complicated. Indeed, it

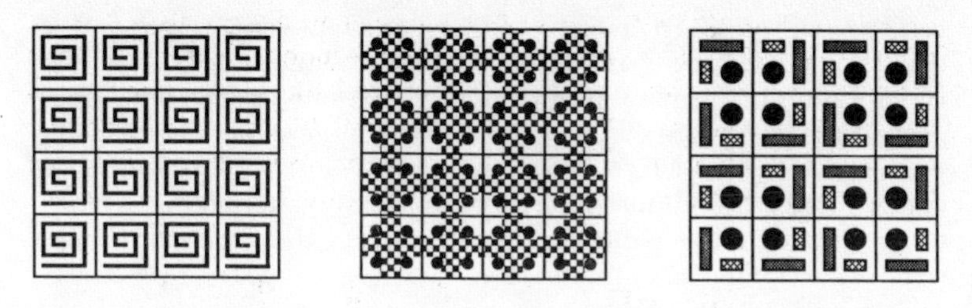

Figure 9.9 Three different patterns based on a square lattice

Figure 9.10 Colour symmetry in an Escher lithograph

can be something of a challenge to look at a repeating design and to decide what all the symmetries of that design actually are. The classification of planar repeating patterns seems first to have been completed by Fedorov in 1891 and made more generally available through the work of P. Niggli and George Pólya in 1924. The result is that there are 17 different types of repeating patterns: the symmetry groups of these patterns are often called the *wallpaper* groups (Figure 9.11). Similarly, there are 46 two-colour repeating patterns.

Of course, it's possible, if you're very careful, to work out by hand the repeating patterns associated with the 17 wallpaper groups. This was done by the Moors, for example, as designs based on each of these symmetry groups have reputedly been found in the Alhambra, a group of buildings overlooking the city of Granada, that were built between the years 1230 and 1354 at the height of the Moorish culture in Spain. The walls and ceilings of these buildings are decorated with intricate geometric ornamentation that exhibit these patterns – and many others, including ingenious approximations to patterns that, if exact, would be impossible. But it's more difficult to *prove* that these are all of the patterns: that's a task best left to the mathematicians.

With all of this cultural and scientific attention to repeating designs in mind, it should come as no surprise that one morning, while eating breakfast in the dining room of a small hotel in Minneapolis, we should be treated to a human interest story about how textile patterns are made. We had no choice but to pay attention, because 'continental' breakfast – as many varieties of doughnut as there are wallpaper groups – was included in the cost of the room. The management had also thoughtfully provided a huge television screen, smack in the middle of the main wall, with a powerful sound system. Although we tried to avoid the invasion of our airwaves by this ungodly and intrusive means of communication, we failed, and eventually succumbed to its soothingly hypnotic spell. Jolted to our senses by the next commercial break, we realized that it would be possible to design quilts using chaos; and the results have been most agreeable.

Recall how chaotic dynamics in the plane is created. We choose a mapping $f(x)$ from the plane to itself, and an initial point x_1. We form the infinite sequence $x_1, x_2, \ldots x_n, \ldots$ by iteration; that is, $x_{n+1} = f(x_n)$. If all the points in this sequence are the same, then we have a *fixed* point. If alternate points are the same, we have a *period two* point. Loosely speaking, if the points in our sequence stay bounded and yet seem to wander around without a discernible pattern, then we call the dynamics *chaotic*. We shall call the collection of all the points in such a chaotic sequence a *chaotic set*.

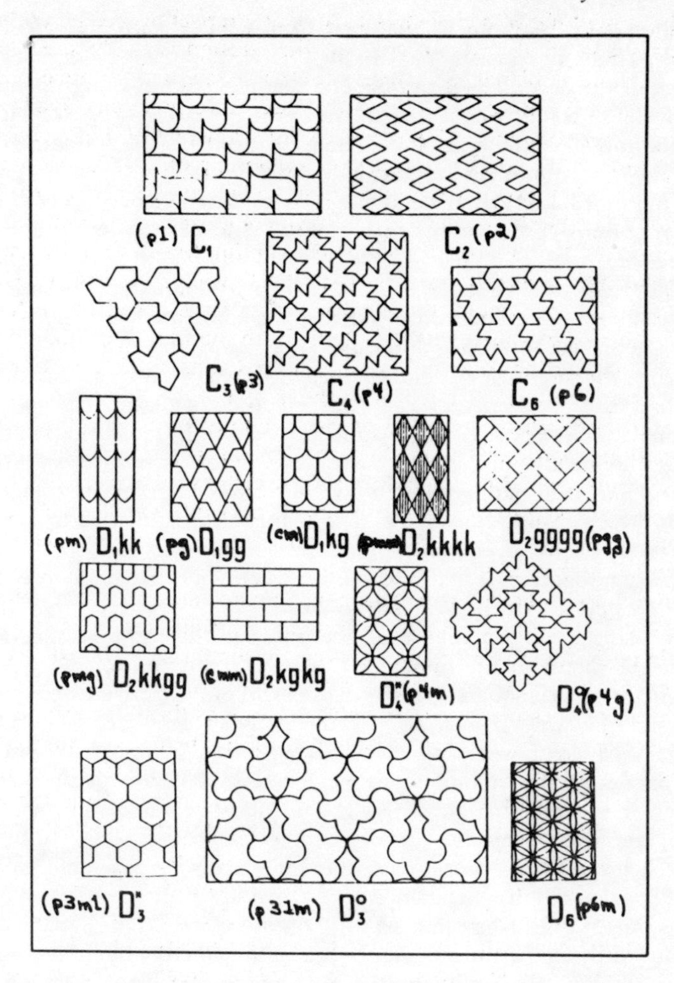

Figure 9.11 Pólya's original drawing of the seventeen wallpaper patterns. He has added the international crystallographic symbols by hand

Suppose we want to make a design that repeats periodically on the square lattice, which is generated from a square fundamental cell by horizontal and vertical translations through equal distances. To create a repeating pattern we need to choose a mapping that preserves the square lattice. That is, suppose we take a point in the plane and another point that can be obtained from it by a translation in the lattice. Then, for consistency with the lattice, we must require that the image of the second point under the mapping can also be obtained from that of the first by a translation in the lattice. Effectively, we're identifying all of the cells in the square lattice with the fundamental cell, the unit square, and requiring the mapping to be consistent with that identification. When we iterate the mapping, we need only write down our answer in the fundamental cell; and that gives a bounded sequence.

The process we've just described always leads to a set (a pattern) that can be repeated periodically across the plane, matching perfectly at the joins. In the peace and quiet of a hotel room we decided to enhance the symmetry characteristics of this pattern by demanding that on the fundamental cell, the mapping actually has square symmetry. We worked out a formula for the simplest mappings of the plane that meet all of these requirements – if you want to know what it looks like, it's in appendix 2, along with two computer programs to draw the corresponding quilts. The main thing you need to know here is that it involves four arbitrary real numbers a, b, c, d, and an arbitrary integer k. To find interesting quilts, you have to choose the right numbers. Which? We had no idea.

So, using a lap-top computer, we typed in the formula, guessed plausible values of the numbers a, b, c, d, k, hoping to find a few chaotic sets with square symmetry – and to our surprise discovered one straight away. The reason for this early success became rapidly clear: nearly every set of numbers that we tried ... worked. We show the results of this iteration process for several values of the parameters in this map in figure 9.12. The results are complex, intriguing, and arguably beautiful. Some – ironically, given our 'hi-tech' approach – look like floral patterns; some strongly resemble batik; and one looked like a stained glass window influenced by both Christian and Muslim images. All of these varied designs were produced by the identical, simple procedure, merely by adjusting a few numbers. It's worth noting that producing black and white pictures on a computer is rather an easy task – producing the high-resolution colour ones shown here is much more difficult. The colourings on these pictures were produced by Mike Field.

The results have two important implications, over and above their aesthetic appeal. The first is that symmetry and chaos – pattern and disorder – can coexist naturally within the same simple mathematical framework. The second is that the chaotic patterns produced by this technique *look* complicated, yet they are prescribed by a short computer program and a few numbers. Their information content is

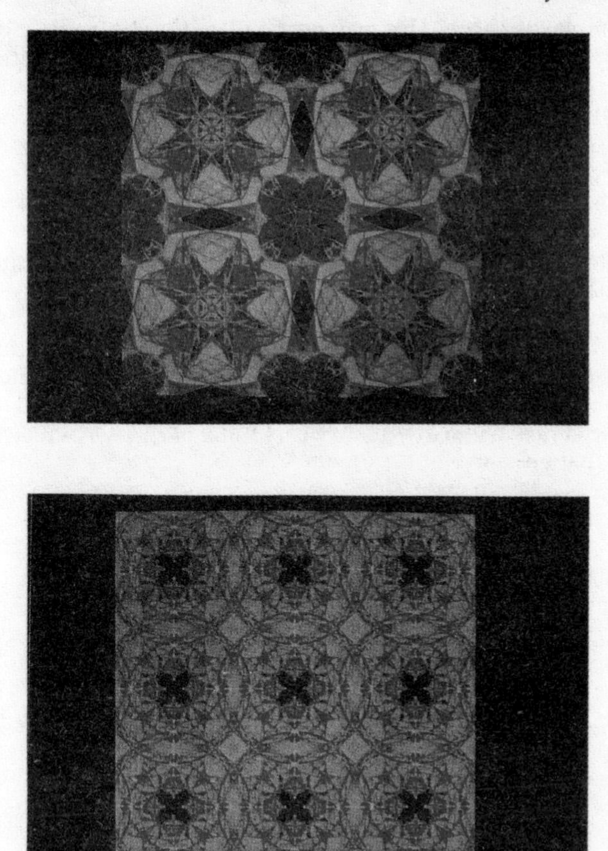

Figure 9.12 Quilt patterns designed by symmetric chaos

thus actually very small. Normally we think that it takes a lot of information to specify a complicated structure – but here it doesn't. There are two possibilities. One is that complexity is not conserved: it can be created from nothing. The other is that forms that appear to be complex may not actually be so. The latter is, to us, a more appealing interpretation: the quantity of information needed to *des*cribe an object may be more than that needed to *pre*scribe it. But it implies that you can't tell how complex something is just by looking at it; you have to consider all possible processes that might generate it, and ask how complex *those* are. This may not be easy. So a by-product of our messing about with quilts is a serious question about the nature of complexity. Other mathematicians and physicists have raised the same question from different points of view. In particular Gregory Chaitin has developed a theory of randomness based upon the difference between the information needed to describe a sequence of digits and the information needed to prescribe it. He calls it Algorithmic Information Theory.

The investigations that we've just related appear to be the tip of a mathematical iceberg, the theory of symmetric chaos. It poses a whole new range of dynamical questions, and the answers so far obtained include several surprises. Moreover, it's extremely curious that the intervention of chaos, and symmetry-on-the-average, leads to symmetry-creation as the natural phenomenon, rather than symmetry-breaking. We don't normally think of chaos as a means for creating pattern.

10

Well, Is She?

Professor Jones had been working on time theory for many years.
'And I have found the key equation,' he told his daughter one day. 'Time is a field. This machine I have made can manipulate, even reverse, that field.'
Pushing a button as he spoke, he said, 'This should make time run backward run time make should this,' said he, spoke he as button a pushing.
'Field that, reverse even, manipulate can made have I machine this. Field a is time.'
Day one daughter his told he, 'Equation key the found have I and.'
Years many for theory time on working been had Jones Professor.

Frederic Brown, *Nightmares and Geezenstacks*

To Plato and the Pythagoreans, the universe was governed by fundamental mathematical patterns. The statement 'God ever geometrizes', our motto for this book, is attributed to Plato; and it's also said that over the door of his academy was a sign reading 'Let no one ignorant of geometry enter.' Out of deference to classical scholarship, we must add that these quotes are probably paraphrases: the closest we've been able to document is in the *Gorgias*, where Plato says that 'geometric equality is of great importance among gods and men'. The Platonic philosophy, and even more so the Pythagorean, saw numbers and geometric forms everywhere – often when they weren't actually present. There's a great deal of numerology and mysticism in the Platonic world-view. A sample of the genre, from *The Republic*:

> And if a person tells the measure of the interval which separates the king from the tyrant in truth of pleasure, he will find him, when the multiplication is completed, living 729 times more pleasantly, and the tyrant more painfully by this same interval.

Even today, the mysticism hasn't vanished entirely. The crystallographer Alan MacKay, in an article called 'But What is Symmetry?', remarked that

> We have a Pythagorean strain in our culture which has continually made congenial the idea that somehow the symmetrical geometrical figures – the five Platonic solids in particular – are at the bottom of things.

But, as he explains in the same article, these traces of Platonic mysticism survive because we find them useful:

> Discourse about solid structures is impossible without effectively being able to call up pre-fabricated concepts, level upon level, the simplest being the Platonic solids, as we will. Literary labels, such as the words 'rhombic triacontahedron' or 'para di-chloro-benzene' have precise meanings. If we do not know enough of them, then we cannot even begin to use the hierarchically structured tree of concepts which is modern science.

In previous chapters, our adherence to Platonic principles has led us to a variety of mathematical concepts, related to geometry and symmetry, and applications of those concepts to science. The time has come to take stock. What have we learned, why is it worth knowing, what are its limitations, and what does it all mean? *Is* God a geometer?

Platonic Relationship

Lynn Arthur Steen has described mathematics as 'The science of pattern'. Mathematicians, perhaps more than other scientists, appreciate the power of logical structure. They don't mind pursuing abstract generalities, if necessary for decades, before any practical pay-off appears. Some have no wish to be involved in practical pay-off at all, preferring to follow their noses and do whatever seems most interesting; others like to take new ideas and develop them to the point where they can be used in science, industry, or commerce. Still others like to start from problems in science, and see what kind of mathematics these demand. Advocates of one or other of these sub-philosophies miss a crucial point: mathematics advances most rapidly when all of these activities are pursued simultaneously.

Mathematicians work in this diversified manner because they find it natural, because they enjoy it, but also because they think that it's the best way to get results. The working philosophy of most mathematicians is, in practice, Platonic: they believe that archetypal structures, fundamental patterns, have their own kind of existence, and talk of 'discovering' them rather than 'inventing' them. If pushed into an intellectual corner, they'll generally agree that it can't be as simple as that; nevertheless, they operate as if they are exploring a new continent, rather than making everything up as they go along.

That philosophy, odd though it may seem when phrased that way, has paid off: our deepest insights into the natural world have been mathematical, and the idea that the laws of nature are couched in mathematical terms has become a cornerstone of modern science. In 1610 Galileo said that the language of nature is mathematics: 'its characters are triangles, circles, and other geometrical figures'. In 1939 Dirac wrote:

> The mathematician plays a game in which he himself invents the rules while the physicist plays a game in which the rules are provided by Nature, but as time goes on it becomes increasingly evident that the rules which the mathematician finds interesting are the same as those which Nature has chosen.

More recently Eugene Wigner (see Further Reading) puzzled about the 'remarkable effectiveness of mathematics' in describing the physical world.

All three echo Plato's assertion that 'God ever geometrizes', with which we began our quest. In *Fearful Symmetry* we've taken Plato literally, and focussed upon our own particular, rather narrow, interpretation of 'geometry'. Our central argument has been that Plato's assertion is justified – not as a statement about a personal deity, but as a metaphorical way of expressing the existence of many of the regular patterns in the world around us. We've concentrated on just one aspect of that viewpoint, the idea of symmetry-breaking, which unifies many processes of pattern-formation. From a sufficiently deep and general viewpoint, geometry and symmetry are synonymous. According to Felix Klein, each type of geometry is the study of the invariants of a group of transformations; that is, the symmetries of some chosen space. Symmetries, in turn are captured by group theory. So, if Plato and Klein are correct, then God must be a group-theorist.

Is She?

We've amassed quite a lot of evidence in support of that position. Symmetry in nature is varied and widespread. A substantial number of the patterns that occur in the world can be described and analysed in a unified manner through the concept of symmetry-breaking. They range from everyday items such as a crushed coke can to the deepest principles that govern the structure of space, time, and matter. The God of Crystals knew about the 230 crystallographic groups, and employed them in the construction of mountain ranges and desert sands, over a billion years before crystallographers classified them and proved their list complete. The Goddess of Evolution breathed translational symmetry into arthropods a billion years before She breathed bilateral symmetry into crystallographers, thereby enabling that act of classification.

However, there are two columns on every balance sheet.

Square Orbits

It's all too easy to see patterns where none exist, or to force things into predetermined patterns whether or not they fit. Kepler was immensely proud of a curious theory of the distances between planets, which he related to the regular solids. He imagined the planets to be placed upon a series of concentric spheres, with the solids sandwiched tightly in between (Figure 10.1). To him it was no idle speculation: in the preface to his *Mystery of the Cosmos* of 1596 he makes it the centrepiece of his work:

> I undertake to prove that God, in creating the universe and regulating the order of the cosmos, had in view the five regular bodies of geometry as known since the days of Pythagoras and Plato, and that he has fixed according to those dimensions, the number of heavens, their proportions, and the relations of their movements.

It's just such an attitude that Jonathan Swift ridicules when relating the experiences of Lemuel Gulliver on the flying island of Laputa:

> Their ideas are perpetually conversant in lines and figures. If they would, for example, praise the beauty of a woman, or any other animal, they describe it by rhombs, circles, parallelograms, ellipses, and other geometrical terms. I observed in the king's kitchen all sorts of mathematical and musical instruments, after the figures of which they cut up the joynts that were served to his Majesty's table.

It is, however, worth bearing in mind the words of István and Magdolna Hargittai, which relate Kepler's outmoded planetary model to his epoch-making work on crystal structure, which we described in chapter 4:

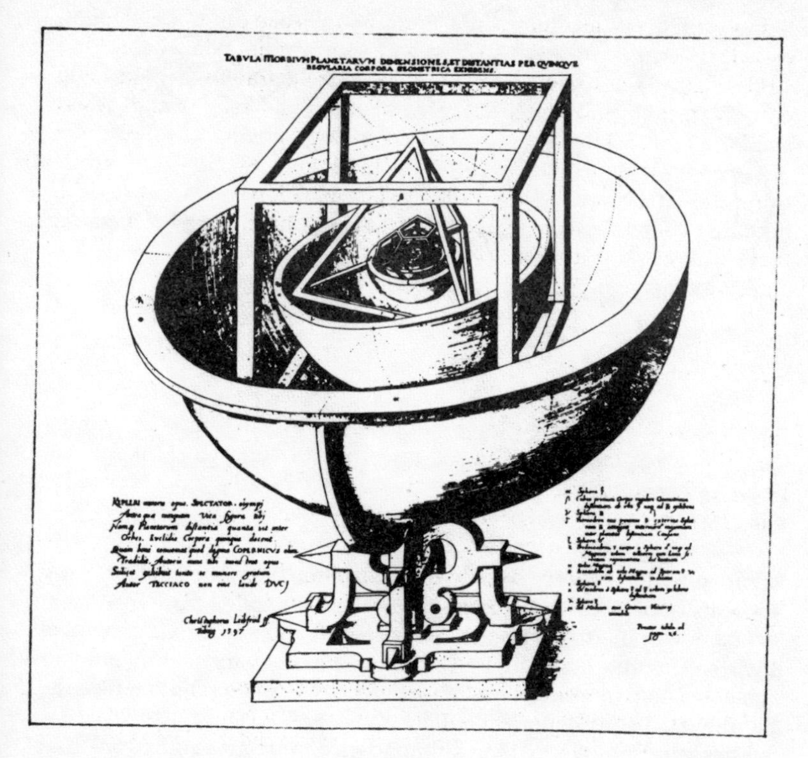

Figure 10.1 Kepler's theory of planetary distances: regular solids sandwiched between celestial spheres

> Arthur Koestler in 'The Sleepwalkers' called this planetary model 'Kepler's most spectacular failure'. However, the planetary model which is also a densest packing model probably symbolizes Kepler's best attempt at attaining a unified view of his work both in astronomy and what we would call today crystallography.

Instead of imposing nonexistent structure, it's also far too easy to impose a structure that's so flexible that it can be made to fit any observations whatsoever. Plato and Eudoxus devised a system of epicycles (circular orbits about points moving in circular orbits about points moving in ...), and Ptolemy developed it by adjusting the sizes and speeds of the component circles, until it gave an exquisitely accurate fit to the motions of the planets (Figure 10.2a). However, it

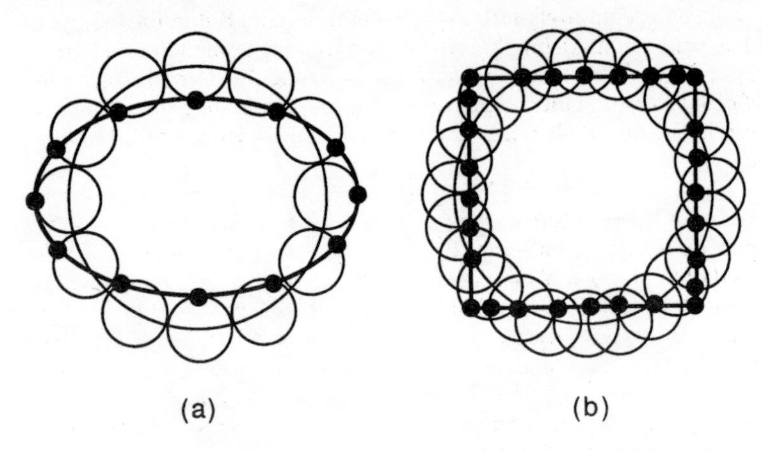

Figure 10.2 Ptolemy's epicycles can (a) model planetary orbits, and (b) model anything else too

could also have been adjusted for an equally exquisite fit to any motions whatsoever. Square orbits, for example, pose no great problem within Ptolemy's general system (Figure 10.2b). Just revolve the smaller circle in the figure – the speed has to vary slightly but you can avoid that by using even more epicycles – and a point on its edge happily traces out a perfect square.

These are the two extremes that science must avoid: theories that don't agree with reality, and theories that can be made to agree with anything. However, the latter do have a place as empirical descriptions: Plato's epicyclic model was *useful*. It predicted the positions of the planets. But they're seldom the last word: they're superficial, little better than tables of observations cast into a convenient form. They can generally be recognized by a proliferation of 'adjustable parameters' – numbers that the theorist can choose, after the event, to improve agreement with what actually happened. The *interesting* patterns, the ones that scientists consider as deep and fundamental, are those that leave little room for that kind of manoeuvre. The patterns of crystal lattice symmetry, for example, aren't adjustable at all. Mathematics presents us with a complete, final list. If the structures on that list didn't occur in the natural world, we'd have to discard the whole idea. As it happens, in this case mathematics and nature agree.

The Platonic viewpoint has misled scientists on occasion, usually during the early development of theories, when detailed structures have been prematurely elaborated by enthusiasts. But in the long run it has paid off handsomely. We've seen that the somewhat maligned 'crystallographers' were *right* to imagine that the forms of crystals reflected mathematical structure, even though they made every mistake in the book when they tried to guess what that structure might be.

Indeed, despite Swift's biting satire, scientists and mathematicians aren't the main culprits among those who place too heavy reliance upon non-existent patterns. It's a common human failing. A whole breed of financial analysts currently attempts to predict the behaviour of the stockmarket by applying a range of 'patterns', either geometrical or numerological, whose basis is – to say the least – dubious. The sensible ones make money by teaching the system to everybody else. Several schools of architecture – often respectable and respected – are based upon number mysticism. Le Corbusier's 'modulor' emphasizes ratios based on Fibonacci numbers and the golden ratio. It's not that people don't design good buildings by these methods; it's that their design sense plays by far the greatest role, and the mystical framework is so flexible that any reasonable design can be incorporated into it. Meanwhile naive and impressionable people think something clever and deep is going on. Pseudoscientists, from pyramidologists to flat-earthers, promote systems of alleged pattern that either bear no resemblance to reality or resemble it by virtue of total flexibility.

Production-line Universe

We've talked a lot about how symmetry breaks – but we've said very little about where it comes from in the first place. It's time we remedied that omission.

Some symmetries are imposed by human agency. We can *make* a spherical ping-pong ball or a cylindrical coke can. We usually do. Their symmetry is convenient for technological control and development; but we could put fizzy drinks into asymmetric cans if we wanted to. For some reason, we seem to prefer symmetrical things. Aeronautical engineers have done calculations showing that an aircraft with one wing swept back and the other swept forwards actually has advantages over the conventional bilaterally symmetric configuration. However, no aircraft manufacturer has yet dared to make a jumbo-jet with a skewed wing: it's unlikely that the public would trust such an ungainly design.

While we may not understand the reasons for this human preference for symmetric objects, it's easy to see how that could lead to their widespread manufacture. Manufacturing process themselves are conducive to symmetry: it's easier to make lots of copies of the same thing. Human-inspired symmetry isn't such a problem. But symmetries arise in nature, spontaneously, and pose much deeper questions. Raindrops and planets are spherical. Crystals have lattice symmetries. Galaxies are spiral. Waves on the ocean are spaced periodically. The polio virus is an icosahedron. Hornets, hamsters, harriers, herrings, and humans are bilaterally symmetric. The methane molecule is a tetrahedron, with a carbon atom at the centre and a hydrogen atom at each of the four vertices. A new form of the element carbon, known as 'Buckminsterfullerene' after the famous architect, has recently been discovered: it has 60 carbon atoms, arranged at the vertices of a truncated icosahedron (Figure 10.3). Although designed to human specification, it's proving unusually stable, and is believed to exist naturally in interstellar space.

How does all this symmetry arise?

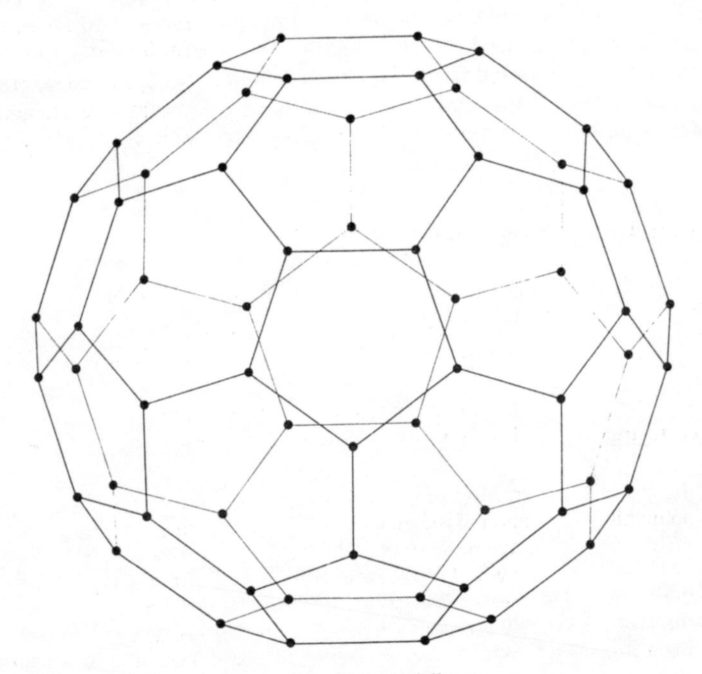

Figure 10.3 Molecular structure of Buckminsterfullerene

The common underlying cause seems to be the fact (itself an even deeper mystery) that the universe is 'mass-produced' – it's composed of large numbers of identical bits, rolling off the cosmic production line. Every electron looks like every other electron, every proton like every other proton, every charmed quark like every other charmed quark. (Richard Feynman wondered if that was because there was really only *one* electron, wandering backwards and forwards in time.) The laws of nature, moreover, appear to be the same everywhere – though we may yet discover that, like so many human assumptions of homogeneity, this isn't so. We can therefore argue much as we did when discussing crystal symmetry. Mathematically, the equations that describe a system composed of a large number of identical components have a vast amount of symmetry. They're symmetric under all possible permutations of the components, for example, not to mention rigid motions of space and time. It's reasonable to suppose that the actual symmetry that we observe in nature is a broken version of this 'potential' symmetry present in the mathematics.

This kind of reasoning applies to viruses, for example, which are 'compact crystals'. In order to infect a cell, they have to be small, so they can't form lattices. But, like crystals, they're built from a number of identical components. Principles such as the minimization of energy, which implies stability of structure, can then lead, quite naturally, to arrangements with polyhedral symmetry.

Dicing Deity Meets Geometer God

There's one sense in which a physical system, and not just a mathematical ideal, may also possess symmetry under all possible permutations of its components. Recall that a chaotic system can have an attractor with a considerable degree of symmetry, and that this manifests itself physically as 'symmetry in the mean'. Let's apply this idea to the molecules in a gas, following Henning Genz:

> To get at the interesting properties of a gas, one must take averages; something the motion of the molecules does all the time. This way, one arrives at the average number of molecules in any given volume of the gas. If this density is independent of the position of the volume, then the gas is 'in the mean' translation symmetric. Thus by averaging over the positions of the individual molecules one arrives at the complete motion symmetry of the gas at the macroscopic level. In this sense chaos and symmetry are equivalent.

> Crystals are a symbol of symmetry and yet their order breaks the complete macroscopic symmetry of a melt or gas. This is a general rule: by ordering finite parts it is impossible to reach the complete symmetry in the mean resulting from chaos. Consider a television set that has not been turned off after the end of the program. The screen then flickers chaotically and is therefore (except at the boundaries) in the mean completely motion symmetric. Suppose now the flickering stops, and one of the typical translation and rotation symmetric drawings of M. C. Escher appears on the screen. This would reduce the previously prevailing symmetry under *arbitrary* rotations and translations to a symmetry under only *certain* translations and rotations.

His remark about averaging chaotic motions to obtain symmetry resonates with chapter 9, and also explains why we were able to ignore the atomic irregularities of an allegedly symmetric pond in chapter 1. Because of symmetry on the average, bulk matter – which on very fine scales has detailed atomic structure – can sensibly be represented as being homogeneous or isotropic in traditional continuum models.

Theory of Everything

The extent to which symmetries pervade the universe is remarkable. We've seen examples of symmetry-breaking, on scales that range from molecular vibrations to intergalactic voids. If current theories at the frontiers of physics are correct, broken symmetry lies behind the most fundamental small-scale phenomena: the interactions of elementary particles. You may recall being told that atoms are built up from a nucleus, composed of two types of particle known as the proton and the neutron, and that these are orbited by a mass of smaller particles known as electrons. However, there have at various times been more than sixty different kinds of 'elementary' particle, so it's no longer that simple!

As we mentioned briefly in chapter 7, the interactions of fundamental particles involve four distinct forces. Each force is thought to be the consequence of interactions in which one or more types of particle are exchanged. The standard analogy is a game of tennis: the continual exchange of the ball from one player to the other keeps both of them within the confines of the court while the game lasts, producing an effect of attraction.

In order, from the weakest to the strongest, the fundamental forces are:

- *Gravity* Has the longest range, and acts on all matter. Involved in the orbital motion of planets. Exchange particle: the (conjectured) graviton.
- *Weak nuclear force* Very short range, acts on everything except photons – particles of light. Involved in beta decay of neutrons. Exchange particles: the W and Z boson.
- *Electromagnetism* Long range, acts only on particles possessing electric charge. Involved in chemical reactions. Exchange particle: the photon.
- *Strong nuclear force* Short range, acts on quarks and gluons. Involved in nuclear reactions. Exchange particles: gluons.

This is a strange rag-bag of rather arbitrary properties, but physicists are convinced that there must be some simple underlying rationale (once more the cultural bias towards Pythagoreanism?). There are very successful theories that explain how the individual forces work. On the classical (nonquantum) level, Albert Einstein's theory of General Relativity deals with gravity, and James Clerk Maxwell's laws handle electromagnetism. Max Planck introduced quantum notions into electromagnetism, endowing light with a wave-particle duality. The weak nuclear force was discovered in the 1930s, the strong nuclear force in the 1970s; both are described by quantum mechanics.

In Einstein's day only gravity and electromagnetism – relativity and quantum mechanics – were recognized. He expended a great deal of effort in the search for a theory that would unify them both; but even today nobody has succeeded in finding a convincing theory of quantum gravity. However, the electromagnetic and weak nuclear forces now exist within a unified theory, the electroweak theory, based upon quantum electrodynamics (QED). It was confirmed by the discovery of three conjectured exchange particles, the W^+, W^-, and Z^{0+} bosons. The strong nuclear force is described by a theory analogous to QED, known as Quantum Chromodynamics (QCD). In QCD the familiar protons and neutrons are not indivisible: they're built from particles called *quarks*, and held together by *gluons*.

The discovery of quarks is a fascinating story: we record a few highlights. In 1962 Murray Gell-Mann and Yuval Ne'eman discovered that the class of particles known as hadrons – including the proton and the neutron, but also many more exotic particles such as pions, kaons, and lambda and sigma particles – has a beautiful internal structure. When ordered by various quantum properties, such as spin, charge, and strangeness, hadrons display themselves in geometric patterns (Figure 10.4). Underlying these patterns is a symmetry group, known as SU(3). Using the SU(3) symmetry, for example, a

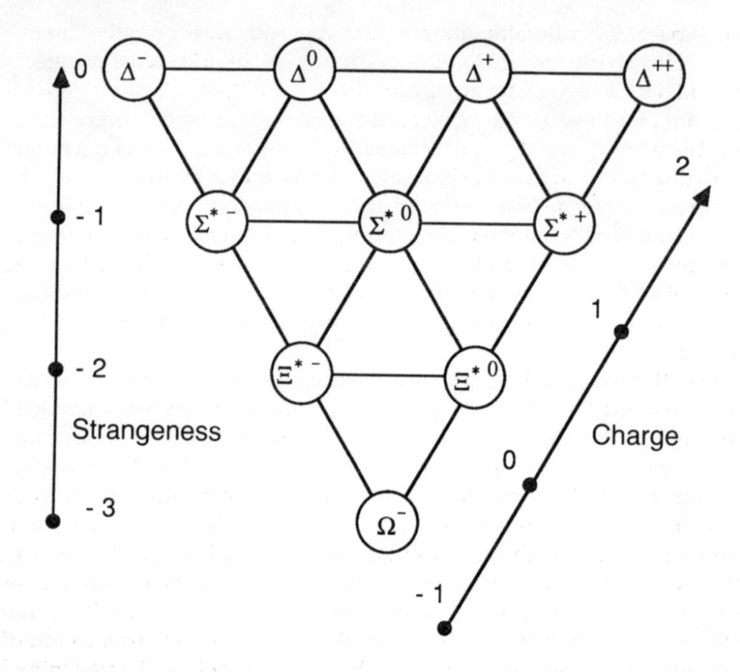

Figure 10.4 Geometric patterns of quantum properties of fundamental particles

proton can be mathematically 'rotated' to yield a neutron. This group is analogous to the rotation group SO(3) in three-dimensional space, but the coordinates of the space must become complex numbers – involving the square root of minus one – instead of being real.

In 1964 the SU(3) theory was confirmed by the discovery of a 'missing' particle that it predicted, the Ω^- (omega-minus). But what gave rise to the SU(3) symmetry? Gell-Mann and George Zweig suggested that all hadrons are composed of the same basic subparticles – named 'quarks' after an obscure passage in James Joyce – and that the symmetries are effectively rearrangements of constituent quarks. Basically, it's the same kind of idea that the crystallographers had: crystal symmetries must occur because crystals are built from lots of identical components which can be swapped around. But quarks had to have weird properties – charges either $\frac{2}{3}$ or $-\frac{1}{3}$ that of the electron, for example – so at first the theory was not received especially enthusiastically.

As particle accelerators became more powerful, experimental support grew, and by 1970 the existence of quarks was generally considered to be established. The theory was then elaborated, with the aim of bringing ever more particles under its sway. The upshot of all this activity was QCD. The electroweak theory followed when it was discovered that at sufficiently high energies of interaction, the electromagnetic and weak forces effectively merge. At high energies, there's an extra symmetry between these two forces, which breaks at the lower energies typical of the world in which we live. A Platonic view of this would be that there exists an 'ideal' universe in which the two forces are on an equal, symmetrically related footing – but *we don't live in that one.*

Has the Geometer God bungled our universe?

At any rate, unification of two fundamental forces was achieved through symmetry-breaking. (Depending upon your viewpoint, this may be thought of as either spontaneous or induced symmetry-breaking – spontaneous if you include the 'ideal' universe in your model, induced if you stay within the universe that we inhabit – and physicists work with both approaches.) QCD plus the electroweak theory provides the so-called 'standard model', the current state of the art in particle physics. However, successful though it is, the standard model doesn't achieve the goal of a complete unification of all four forces; and physicists are hot on the trail of Grand Unified Theories (GUTs). The enormous disparity between the four forces, and thus between their characteristic particles, is a serious obstacle to any symmetry-based theory, because symmetry implies interchangeability. However, there seems to be a way out. If the interactions take place at even higher energies, then the strengths and ranges of the forces change. At an energy around 10^{12} times as high as that required to merge the electromagnetic and weak forces, it is believed that the strong force also merges. The appropriate symmetry group is SU(5), all 'rotations' in a five-dimensional complex space. Unfortunately such energies are well beyond what can be achieved in particle accelerators, so direct confirmation of the SU(5) theory appears out of reach. However, the theory is very attractive on grounds of mathematical elegance.

The odd force out is gravity. What's needed is a fully unified 'Theory of Everything'. The current favourite is the theory of superstrings. This considers particles not as isolated points in space, but as extended multidimensional closed surfaces, or 'strings'. Strings can vibrate, giving rise to various quantum properties; they can merge or separate, like interacting particles; and their geometric nature lends itself to being made relativistic. Currently, the most favoured symmetry groups in superstring theory are SU(32), 'rotations' in 32-dimensional complex space, and a more exotic creature called $E_8 \times E_8$. String theory operates best in spaces of higher dimension than the familiar four of time and space. The idea is that the extra dimensions are curled up tightly into balls no more than 10^{-35} metres across, so that we don't ordinarily notice them. When the physicist Steven Weinberg lectured on this theory, Howard Georgi, in the audience, composed a limerick to celebrate the idea:

> Steve Weinberg, returning from Texas
> brings dimensions galore to perplex us.
> But the extra ones all
> are rolled up in a ball
> so tiny it never affects us.

Particle physicists keep changing their theories. As we write, claims that there exists a new type of neutrino, unexplained by anything we've so far described, are being hotly debated. As Georgi also says: 'My personal belief is that Nature is much more imaginative than we are.' But all of these changing theories have a common message – that the structure of fundamental particles, the basis of our entire universe, is controlled by broken symmetry.

Nature or Nurture?

All of this leads us to assume that the mathematical patterns that scientists observe in nature are not delusions: they're really there. The question is, to what extent are they fundamental to the way nature works; and to what extent are they just convenient descriptions that the human mind can grasp?

The symmetry groups of crystal lattices, for example, are without doubt fundamental to the human understanding and exploitation of crystals. But did the God of Crystals *really* use the lattices to assemble His creations? Does Nature 'know about' crystal lattices? Probably not. The basic pattern involved is that matter is composed of large quantities of identical pieces – atoms of a given element, molecules of a given compound – and that those pieces interact according to common physical principles. The lattice structure is (presumed to be) a *consequence* of basic atomic physics. The God of Crystals didn't have to invent lattices: they happened by themselves.

Moreover, lattices only occur in very special circumstances: slow growth under uniform conditions. We've seen that so diverse are the actual forms that minerals assume, that for centuries there was a serious dispute whether there's any true pattern to be found. The pattern is in any case seldom perfect. All crystals possess flaws, some of which are called *disclinations* and others *dislocations*, where the atoms don't fit properly, and where the lattice structure breaks down. Disclinations and dislocations are of great importance in modern science. They are implicated in the fracture of materials under stress, for example; and manufacturers of computer chips would like to avoid them when growing the large crystals of silicon upon which the technology depends. There are *patterns* to disclinations and dislocations, as well (Figure 10.5). From being an unwanted exception to the ideal form they have for many people become the most interesting thing around.

Once we humans have seen a pattern, we tend to become obsessed by it, until we believe that nothing else can possibly be important. Every so often either the real world, or a truly imaginative human being who is usually thinking about something quite different, wakes us up. Quasicrystals, with their forbidden fivefold almost-symmetry, their Penrose pattern structure that is regular but not a lattice, demonstrate that the God of Crystals has a more fertile imagination than we do (Penrose and Kepler excepted!). Benoît Mandelbrot's coinage of the term 'fractal' highlighted a type of geometric structure that had surrounded us, unrecognized, since the dawn of time.

Symmetry is operationally important to human beings because it's an ideal case, one that our brains seem especially able to grasp and to manipulate. Perhaps this is because symmetry reduces information content – it's easier to envisage multiple copies of something simple than a complicated structure of equal size. Perhaps it's because our brains have adapted to cope with nature's symmetries. Our visual cortex, for example, is composed of layer upon layer of cells, with remarkable internal symmetries related to the detection of orientation. During our evolution we needed to be able to recognize a bear, or a rabbit, with equal effectiveness, whether we were standing up, lying on our side, or hanging by our heels from a tree. So our brain has evolved to cope with symmetry; indeed it seems to like it. Why is symmetry aesthetically pleasing? Do symmetric forms produce unusual firing-patterns in our neurons, which somehow leak over into our pleasure-centres? If so, why is too bland a symmetry boring?

Some of nature's symmetries seem to be inherent properties, not merely patterns imposed upon it by the way our brains interpret incoming signals. The fundamental symmetry of space–time – that every time and place obeys the same physical laws – surely holds (or does not) independently of our perceptions. The SU(3) symmetry of

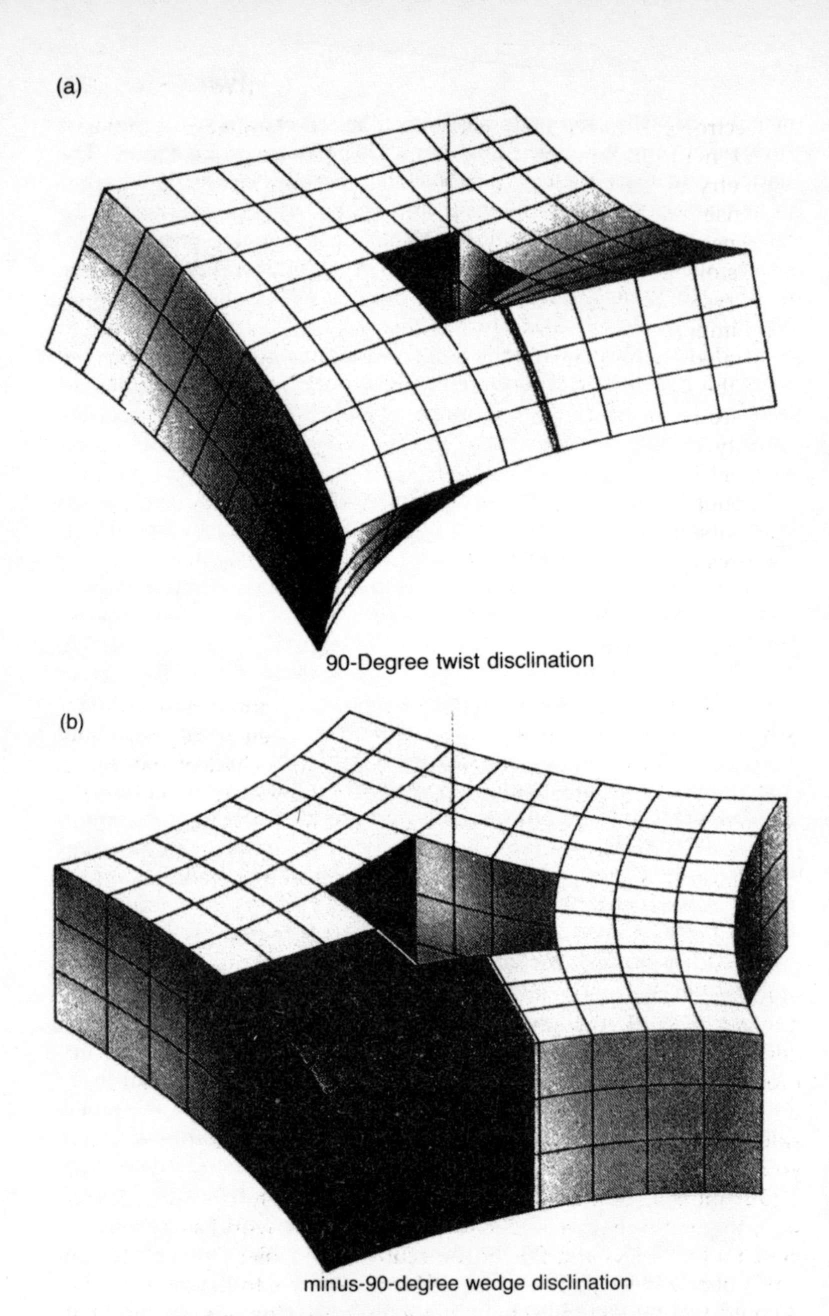

Figure 10.5 Crystal disclinations have their own geometric patterns

the electroweak theory appears to be equally fundamental – but then, so did neutrons and protons, not so long back. Like the lattice symmetry of crystals, it may be merely a human invention – not in the sense that Nature never uses those patterns, for she does; but in the sense that *we* find them a convenient way to deal with some aspects of nature, and we *select* those aspects that fit the pattern. To us, a crystal lattice is 'perfect' and (except to specialists) a dislocation is an 'imperfection', a defect, somehow secondary, something to be denigrated. Nature may well not appreciate that kind of distinction at all, in the following sense. Mathematically, both lattices and dislocations are solutions to the equations that govern the state of a large quantity of atoms: why select one solution as being superior to another? Both lattice and dislocation are concepts imposed by our perception of reality, rather than being inherent features of the way reality itself operates. The God of Crystals may well love all his creatures equally. So may the God of Elementary Particles.

The Human Angle

Are symmetries intrinsic patterns of nature, or artefacts of human perception? There's no universal answer. Some seem to be one, some the other. However, the human brain is itself a child of nature: it evolved in a universe that obeys nature's laws. It wouldn't be surprising if features of those laws are built in to the functioning of the brain. Thus those patterns that the brain is able to detect may not be arbitrary: it may have evolved to detect the patterns that are 'really' present.

Humans evolved to live in a complicated and often hostile world. Our senses are therefore biased towards perceiving particular kinds of things. For example, if the eye detects what it thinks is something moving rapidly towards it, the eyelid shuts automatically. The detection mechanism is crude, but quick. It doesn't wait to work out precisely what, if anything, is approaching the eye: it acts regardless. If you put your face up to a pane of glass and someone on the other side pokes a finger at your eye, you flinch, even though intellectually you *know* it's safe.

Our mental circuitry, in short, has built-in distortions of perception; the world that we see is not necessarily the world as it really is. When a bee looks at a flower, detecting ultra-violet light, it sees all kinds of markings that we miss. A bird, sensitive to magnetic fields, 'knows' what direction to fly when migrating. Humans, lacking that sense, marvel at it; yet a bird can use it as easily as we can hear a muffled noise and decide that the kids have left the TV on again. When primitive tribespeople encountered aeroplanes for the first time, they 'saw' feathers on them. If it flies, it's got to have feathers. Stands to reason, right?

In order to cope with a complex environment, we tend to 'lump' it into convenient categories, reducing the amount of information required to function effectively within it. Innumerable plant species are summed up by the single concept 'tree'; and for many purposes – climbing it to escape from wolves, felling it to make a hut, burning it to keep out the cold – a finer description is unnecessary. This seems to be a very effective trick. One of the reasons for our prevalence as a species is that we're generally better at manipulating the environment to our own ends than other species are. It is of course the same trait that gets us into all kinds of trouble, because we seldom anticipate all the consequences of our manipulations. Our scientific theories structure the universe in the same way: lumping into categories things that *seem to us* to be similar. It doesn't always work, for example 'whale = fish', but we're getting better at refining it, at basing it upon deeper and more fundamental properties. However, our brains lack the capacity to take in the universe as a whole: we have to structure it in order to put little bits of it into our heads. Somewhere in our brains, for instance, is some sort of model of the laws of motion. Imagine yourself tossing a tennis-ball into a nearby wastebasket. You *know*, quite accurately, how hard you'd have to throw it, at which angle, and what trajectory it would follow. Now try it with a real tennis ball, and on the whole you'll find you were right. Sometimes you may even catch yourself trying to influence a real-world event – making a golf ball go down the hole, for instance – by 'thinking' the model of it in your head in the direction you want to occur.

We don't actually have the complete laws of motion in our heads. If you were suddenly transported to the Moon, you'd be in real trouble until you'd learned to build a similar model suited to lunar gravity. In short, the way humans think about the universe is that we *select* or *invent* patterns. They may not be as fundamental as we think they are: but we have very little choice. As a result, however, we may focus our attention on 'basic laws' that are nothing of the kind. The 230 crystallographic groups, for example, are very pretty, but they're far less basic than the laws of quantum mechanics. Quantum mechanics is more general; moreover, it (so we think) *implies* the lattice structures of crystals. So how can we tell which laws really are fundamental?

We can't.

We can't even be sure that the patterns we think we've detected are *laws;* that nature really follows them in every detail. We just have to do the best we can. The problem that we face can be illustrated by John Horton Conway's famous game 'Life'. It's played with a population of blobs that inhabit the cells of an infinite chessboard. At every tick of a notional clock, the configuration of blobs changes to a new one according to very simple rules:

- An empty cell acquires a blob at the next instant of time if it has exactly three blobs as neighbours.
- A blob survives for the next instant of time if it has either two or three neighbouring blobs.

The fascinating feature of 'Life' is that all sorts of self-propagating and mobile structures can occur (Figure 10.6). The 'glider', composed of only five blobs, moves diagonally, twitching its tail as it goes. Slightly larger 'spaceships' do the same kind of thing. Spaceships that are too

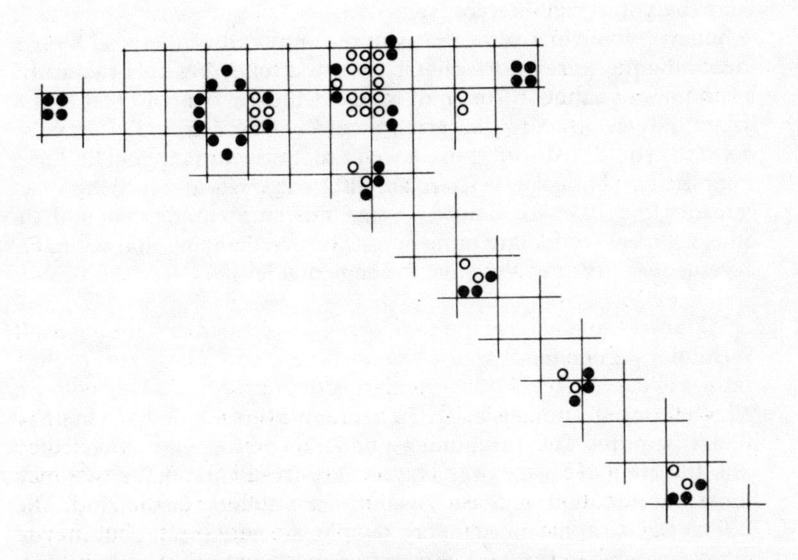

Figure 10.6 Some self-propagating structures in the game of Life. The glider gun (top) emits a stream of gliders (lower right). One snapshot of the process is shown, and for simplicity only every third grid line is drawn. The black blobs will die at the next stage: the white ones will remain alive. New blobs will be born according to the rules of the game. The entire process repeats every 30 generations

big tend to come apart, but can be held together by smaller support vessels if they stay together in a fleet. 'Glider guns' can fire an unending stream of gliders. On a larger scale, streams of gliders can be used like pulses of binary digits in an electronic computer, and an appropriate assemblage of blobs creates a universal Turing machine: a programmable computer. Systems of this type are called *cellular automata*, and they can be made to mimic remarkably complex behaviour. John von Neumann showed that more complicated rules than Conway's can generate self-reproducing systems; Conway showed that his can too. The game of 'Life' generates arbitrary complexity from two trivial rules.

Imagine yourself to be an intelligent 'creature' inhabiting such a cellular universe; a cellular philosopher, or a cellular scientist, who is seeking the fundamental laws that govern its own existence. You'd seize upon the regular 'laws of motion' of gliders and spaceships – or larger objects that were visible to you, gliders and spaceships being analogous to our electrons and photons. You'd get hung up for centuries on the computational patterns of Turing machines. You'd give the 'Nocell Prize' to the genius who discovered the 'genetic code' whereby you reproduce yourself ...

But will you *ever* realize that you're composed of identical blobs, inhabiting the squares of a chessboard, and following only two truly basic 'laws of nature'? The point is, to say the least, moot. And, even if you did get down to that level, would you be satisfied? Certainly not until you'd deduced those laws of motion and that genetic code from the two basic rules! There are different levels at which we may consider the 'laws of nature'. Some are more fundamental than others. But we're fooling ourselves if we ever imagine that we have necessarily arrived at the most fundamental level.

Technology Transfer

Whether or not symmetries are human inventions, they have mathematical features that the human mind can exploit very effectively. One attraction of the concept of symmetry-breaking is that it unifies a huge range of different phenomena. For a given symmetry group (operating in a given manner) there exist universal patterns of symmetry-breaking, which may occur in any physical system that realizes those symmetries. Mathematics is general, and thus is not committed to any specific physical context. This is the advantage of abstraction, and the reason why mathematicians pursue it so avidly, even if the rest of science doesn't always appreciate why.

Symmetries impose such strong restrictions that a great deal can often be said without having any detailed knowledge of the physical equations at all! Many of the most striking phenomena associated with symmetric systems turn out to be *model-independent*, that is, due to the symmetry of the system, rather than to the detailed equations normally used to model it. The detailed equations tell us which of these possible phenomena actually occur; but it's useful to systematize the general effects of symmetry and to develop theoretical methods to handle them, *without* solving the specific equations first. In a sense the approach reverses the classical one, which makes the model equations paramount, and exploits the symmetries only in passing, without special comment or even sometimes awareness that they're present. The idea is to distinguish in advance between what's to be expected and what's surprising. In the Taylor experiment, for example, we've seen that most, if not all, of the bewildering range of flows seen in experiments have a very simple explanation in terms of the cylindrical symmetry of the apparatus. What was once bewildering now becomes completely natural and indeed almost prosaic.

The patterns that occur in Couette–Taylor flow are of interest outside fluid dynamics. For example, if a liquid crystal – such as is used in digital watch displays – is sandwiched in a very narrow gap between two glass plates, and subjected to oscillating electric or magnetic fields, then it forms several kinds of pattern: rolls (which look like stripes), oblique rolls (zigzags), rectangles, and so on. These patterns are very closely related to those for Couette–Taylor flow. Rolls correspond to Taylor vortices, oblique rolls to spirals, rectangles to that curious, usually unstable state of Couette–Taylor flow, ribbons. The symmetries responsible for this resemblance are not *quite* identical to those of the Couette–Taylor system: in fact there's a single additional reflectional symmetry. Nevertheless, being able to follow the paradigm of the Couette–Taylor system makes it much easier to analyse the liquid crystal problem.

Similar patterns have intrigued geologists for many years: they arise when sand or other suspended particles are deposited from relatively shallow water. These patterns are important because they can be detected in rock formations laid down by the sand, and thus give clues to the conditions that prevailed at the time. In this analogy, Taylor vortices become long, unbroken, wave-like sand dunes; wavy vortices are dunes with ripples; spirals are dunes that lie obliquely to the flow of current.

Even the loose analogy that we've noted earlier, between Taylor vortices and the stripes of a tiger, begins to look more plausible at this 'technology transfer' level – especially if you concentrate on the tiger's tail, which is close to cylindrical. The formation of Taylor vortices – fluid stripes – occurs when the perfect cylindrically symmetric flow becomes unstable. In the same way, models of how pigmentation-controlling chemicals might diffuse through the tiger's tail during growth reveal the occurrence of instabilities in the perfect cylindrically symmetric solution – a tail of uniform colour – and these lead to stripes by the identical symmetry-breaking mechanism. Indeed, we've seen that the analogy can be pushed further. In the Taylor experiment, there's a second instability, whereby the vortices themselves break up, acquiring ripples. The analogous effect in the tiger's pigmentation produces spots, and the tiger becomes a leopard.

Taylor vortices, roll patterns in liquid crystals, sand dunes, stripey tigers ... they're just four different real-world manifestations of the identical mathematical phenomenon. (Remember this next time you look at your watch, walk on sand dunes, or visit the tigers at the local safari park.) Their common underlying structure can therefore be understood once, and then used four times. Symmetry-breaking provides a fascinating example of the power of abstract mathematics to facilitate 'technology transfer', in the sense that apparently distinct physical problems may obey identical mathematical rules. Results obtained for one physical realization can therefore be immediately transferred to any other, with little additional effort.

Twist and Shrink

Nature makes use of a whole range of symmetries that we haven't yet mentioned – not because they're less important, but because, for technical reasons, it's harder to make general mathematical statements about them. You should, however, be made aware of their existence: you'll have no trouble finding examples all around you. The simplest of these symmetries, and the only ones we'll discuss, are *dilations*. A dilation is a change of scale: uniform expansion or shrinkage about some specific fixed point, the *centre*. A map is related to the territory that it delineates by a dilation.

Dilations are non-rigid symmetries of space. A finite object can't possess rigorous dilational symmetry, because you can always detect the effect of a dilation by measuring the change in the object's overall size. But infinite objects can possess such symmetry; and finite objects found in nature may well be best modelled as cut-down versions of these. The simplest object with nontrivial dilational symmetry is a cone. Imagine an infinitely large cone, shrinking down to a point at one end and expanding forever like a runaway trombone at the other. If you shrink it to half its size, it looks exactly the same. If you shrink it to one-third, or one-fifth, the size, or double it, or expand it by a factor of a million and two, it also looks exactly the same. The cone is invariant under *all* dilations centred on its tip. Cones also have a rotational symmetry about their axis, and this will prove important in a moment. Conical structures do occur in nature, though not infinitely long ones: the horns of animals, when smooth and evenly tapered, are examples.

How can complete dilational symmetry break? The question is analogous to the breaking of time-translational symmetry that occurs in Hopf bifurcation. Recall that there invariance under *all* translations breaks to yield only invariance under (integer multiples of) *one* translation. Presumably, dilations behave in much the same way. (It's here that our mathematical generalities break down. Hopf's theorem tells us exactly when and how time-translation symmetry breaks; but no analogous proven result for dilations is known. You might not notice its lack, but we do, which is why an otherwise intriguing topic gets only a few paragraphs.) What happens to a cone if its full dilational symmetry breaks, but it remains invariant under (say) shrinkage to half size? The answer is that it develops some structure, such as a bump or a ridge, that appears at various distances along it; with those distances repeatedly doubling towards the open end at infinity, and halving towards the point. A rippled cone, with rather specific conditions on the spacing of the ripples.

A more common shape in nature is the dilational analogue of a rotating wave: a shape that remains unchanged if it's rotated through an arbitrary angle and simultaneously shrunk by an appropriate amount. Shells of many creatures, notably snails, have precisely this form of symmetry. So do the horns of many animals: they come to a point like a cone, but twist in spiral fashion. And not just any spiral: those that have this type of mixed rotational-dilational symmetry are called *logarithmic* spirals.

A further stage of symmetry-breaking is then possible, combining the two already mentioned: logarithmic spiral symmetry, but only through multiples of a fixed angle. For example, the shape might stay the same if you rotate it through a right angle and then halve its size. We might call this *discrete* logarithmic spiral symmetry: it's very common indeed in shells and horns, although the angle need not be a right angle and the dilation need not halve the size. The shell of a chambered Nautilus (Figure 10.7) is an excellent example, and so is the fossil ammonite. Pine cones possess an approximation to discrete logarithmic spiral symmetry – and so do plants that put out side-shoots at regular angles as they grow. Even large trees can possess

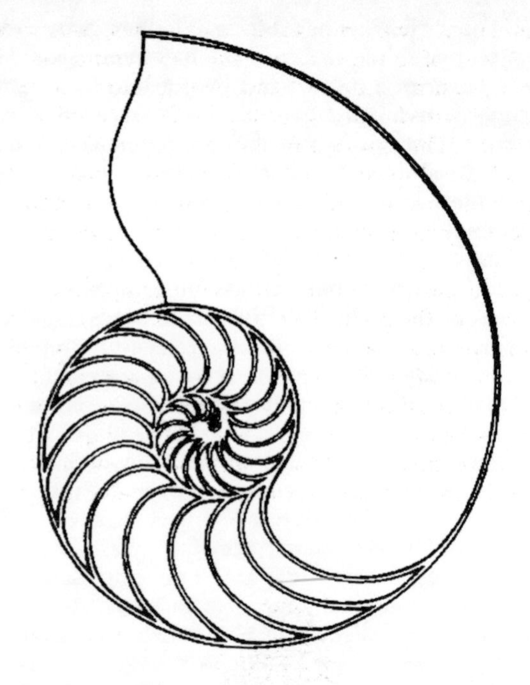

Figure 10.7 Discrete logarithmic spiral symmetry: cross section of a Nautilus

traces of it. So now you'll see trees through new eyes whenever you go for a walk.

Not Just Groups

The group-theoretic paradigm of symmetry has been so successful that it's tempting for mathematicians to assume that it's the last word on the subject. However, the ancient Greeks used the word 'symmetry' in a vaguer and more general sense. MacKay remarks that "Symmetry" is the classical Greek word ΣΥΜ-ΜΕΤΡΙΑ, "the same measure", due proportion. Proportion means equal division and "due" implies that there is some higher moral criterion.' That criterion need not be repetitive structure in the precise group-theoretic sense.

We've encountered various examples that stretch our notion of symmetry. The most striking is the structure of quasicrystals –

'almost-symmetries' with noticeable regularities, but lacking the precise repetition of a lattice. The 'average symmetry' of chaotic systems, the physical trace of a symmetric attractor, is another. Fractals – forms with detailed structure on every scale of magnification – are a third. The simplest fractals are *self-similar* – small pieces of them are identical to the whole, except for a dilation. However, such fractals don't possess dilational symmetry, not in the technical group-theoretic sense, because *several* parts are needed to re-create the whole.

The Sierpiński gasket (Figure 10.8) is the simplest example with which we can make the point. This is obtained by repeatedly deleting the middle quarter of a triangle, removing smaller and smaller pieces, forever. It's named after Waclaw Sierpiński who invented it to show that a curve may cross itself at every point. The Sierpiński gasket can be thought of as being composed of three identical gaskets, each half the size of the original. So it has three 'partial symmetries', three dilations that shrink it towards each corner and halve its size. It is not,

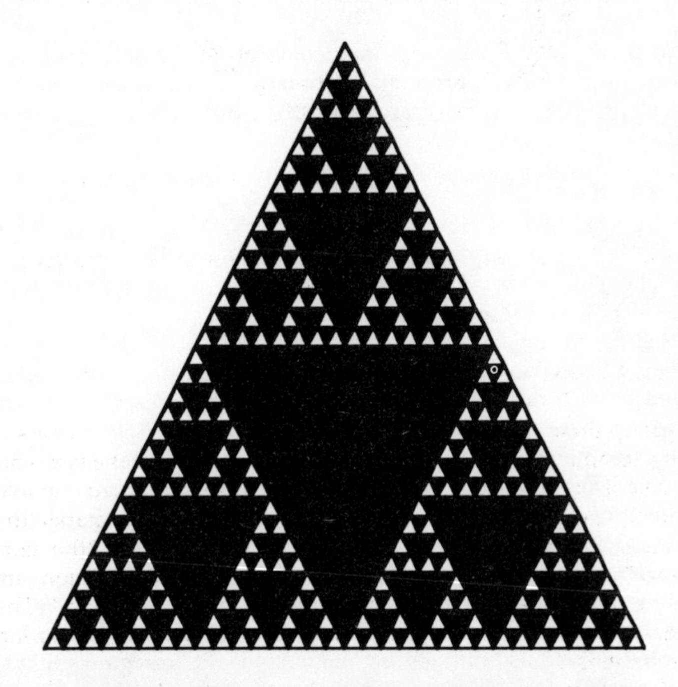

Figure 10.8 The Sierpiński gasket has self-similarities as well as conventional symmetries

however, invariant under any one of these, because each dilation alone produces just a half-sized corner. To reproduce the original shape at full size you need all three copies. So the Sierpiński gasket is invariant not under a single transformation, but under a *set* of three dilations, in the sense that if you apply all three and join the resulting images together, then you reproduce the original. This set of transformations has a clear mathematical structure, but it isn't a group. Michael Barnsley calls it an 'iterated function system'.

Many shapes have partial, incomplete symmetries, or mixtures of different symmetries in different places. Think about a set of regular pentagons arranged in a hexagonal lattice. How much symmetry there is depends upon which aspect of the structure you look at. MacKay notes that:

> Just as Kepler escaped from too rigid a preconceived framework of symmetrical structure for the phenomena of Nature, so also modern crystallographers have gradually generalized their concepts of symmetry, escaping from the rigid formalism of the 230 space groups.

He lists no less than sixty mathematical or quasi-mathematical structures that bear upon crystal symmetry in this wider sense.

In short, God may be a group-theorist, but only on a part-time basis.

Come On, Is She Or Not?

Oh, you still want an answer. It's implicit in everything we've written. Will you settle for that?

We thought not.

We get nervous when people ask for *answers*. We've presented the evidence: we'd rather they formed their own opinions. But, for what it's worth, here's ours. Take it with a large pinch of salt.

The Geometer God is pantheistic, adopting whichever brand of geometry suits the occasion. She uses Euclidean geometry to symmetrize your bodily form, conformal geometry to map your visual senses to your brain, differential geometry to string muscle fibres through your heart, Riemannian geometry to bend the universe and create gravity, symplectic geometry to let there be light.

Or maybe not. Those are all human inventions, things *we* have invented to help us lump natural phenomena into categories that are small enough for our puny minds to encompass. We like *simple* patterns – though we're quick to enlarge our notion of simplicity when a new type comes along that we think we can handle. We refer to dislocations as 'imperfections', yet our Geometer God, who is surely perfect, accepts them without a qualm. Slowly we're learning to distinguish between our limited view of perfection and the much more imaginative interpretation in which we live.

Let's not give ourselves too much credit, though. We didn't invent all of those geometries for ourselves: we stole them. We took hints from the real world. Often the hints arose in quite different areas from the eventual application – Euclidean geometry began as land-measurement and culminated in Newtonian space–time; non-Euclidean geometry began unrecognized as a navigational tool and now underpins General Relativity. But part of Wigner's puzzle about the unreasonable effectiveness of mathematics may have a simpler answer: mathematics is effective in describing the universe because that's where we got it from. However, that doesn't explain the strange manner by which mathematics seems to *amplify* our perceptions. On the one hand we seem to get more out of a mathematical theory than we put in, even though all we do is draw logical conclusions from initial assumptions. Of course, many of these conclusions were not so obvious on a first pass – as anyone who has studied Euclidean geometry knows. On the other hand, mathematics forces us to know *what must follow* from our assumptions. The seemingly simple observation that 'same assumptions' implies 'same conclusions' is a powerful one, and it helps explain why mathematics gives back more than we put in.

In that sense, yes, God *is* a geometer. But never forget: She's much better at it than we are.

Further Reading

R. McN. Alexander and G. Goldspink, *Mechanics and Energetics of Animal Locomotion* (London: Chapman and Hall, 1977)

Wallace Arthur, *Mechanisms of Morphological Evolution* (Chichester: Wiley, 1984)

Wallace Arthur, *Theories of Life: Darwin, Mendel, and Beyond* (Harmondsworth: Penguin Books, 1987)

James Binney and Scott Tremaine, *Galactic Dynamics* (Princeton: Princeton University Press, 1987)

A. C. Bishop, *An Outline of Crystal Morphology*, (London: Hutchinson Scientific and Technical, 1972)

John Tyler Bonner, *Morphogenesis: an Essay on Development* (New York: Atheneum, 1963)

F. J. Budden, *The Fascination of Groups* (Cambridge: Cambridge University Press, 1972)

John G. Burke, *Origins of the Science of Crystals* (Berkeley: University of California Press, 1966)

Jack Cohen, *Living Embryos: an Introduction to the Study of Animal Development*, (Oxford: Pergamon Press, 1967)

Theodore Andrea Cook, *The Curves of Life* (New York: Dover Publications, 1979)

H. S. M. Coxeter, *Introduction to Geometry* (New York: Wiley 1961, 2nd edn 1969)

Paul Davies, *The New Physics* (Cambridge: Cambridge University Press, 1989)

Mike Field and Martin Golubitsky, 'Symmetric Chaos', *Computers in Physics*, 4, 5 (Sept.–Oct. 1990), pp. 470–9

Mike Field and Martin Golubitsky, *Symmetry in Chaos* (Oxford: Oxford University Press, 1992)

David Fowler, *The Mathematics of Plato's Academy: a New Reconstruction* (Oxford: Clarendon Press, 1987)

P. Gambaryan, *How Mammals Run: Anatomical Adaptations* (New York: Wiley, 1974)

Martin Gardner, *The Ambidextrous Universe: Left, Right, and the Fall of Parity* (Harmondsworth: Penguin Books, 1970)

István Hargittai and Magdolna Hargittai, *Symmetry Through the Eyes of a Chemist* (New York: VCH Publishers, 1987)

István Hargittai (ed.), *Symmetry 2: Unifying Human Understanding* (Oxford: Pergamon Press, 1989)

István Hargittai (ed.) *Quasicrystals, Networks, and Molecules of Five-fold Symmetry* (New York: VCH Publishers, 1990)

Gordon Hendricks, *Eadweard Muybridge* (London: Secker & Warburg, 1975)

Edwin Hubble, *The Realm of the Nebulae* (Oxford: Oxford University Press, 1936)

B. Jowett, *The Dialogues of Plato* (5 vols) (Oxford: Clarendon Press, 1875)

Johannes Kepler, *The Six-Cornered Snowflake*, edited and translated by Colin Hardie (Oxford: Oxford University Press, 1976)

D. Kondepudi, 'State Selection in Symmetry-breaking Transitions', *Noise in Nonlinear Dynamical Systems*, (vol.2) ed. Frank Moss and P. V. E. McClintock (Cambridge: Cambridge University Press, 1989)

Helge S. Kragh, *Dirac: a Scientific Biography* (Cambridge: Cambridge University Press, 1990)

J. L. Locher (ed.), *M. C. Escher: His Life and Complete Graphic Work* (New York: Harry N. Abrams Inc., 1982)

E. H. Lockwood and R. H. Macmillan, *Geometric Symmetry* (Cambridge: Cambridge University Press, 1978)

C. H. MacGillavry, *Symmetry Aspects of M. C. Escher's Periodic Drawings* (Utrecht: Bohn, Scheltema, and Holkema, 1976).

J. D. Murray, *Mathematical Biology* (New York: Springer-Verlag, 1989)

R. Penrose, 'Fundamental Asymmetry in Physical, Laws', *Proceedings of Symposia in Pure Mathematics*, 48, pp. 317–8 (Providence: American Mathematical Society, 1988)

F. C. Phillips, *An Introduction to Crystallography* (Edinburgh: Oliver & Boyd, 1971)

Vera C. Rubin and George V. Coyne, *Large-Scale Motions in the Universe* (Princeton: Princeton University Press, 1988)

Doris Schattschneider, *Visions of Symmetry* (New York: Freeman, 1991)

Marjorie Senechal, *Crystalline Symmetries: an Informal Mathematical Introduction* (Bristol: Adam Hilger, 1990)

A. V. Shubnikov and V. B. Koptsik, *Symmetry in Science and Art* (New York: Plenum Press, 1974)

Ian Stewart, *Does God Play Dice?* (Oxford: Basil Blackwell, 1989)

Jean-Louis Tassoul, *Theory of Rotating Stars* (Princeton: Princeton University Press, 1978)

D'Arcy Wentworth Thompson, *On Growth and Form* (2 vols) (Cambridge: Cambridge University Press, 1972)

A. M. Turing, 'The Chemical Basis of Morphogenesis', *Philosophical Transactions of the Royal Society of London*, B237, (1952), pp. 37–72.

Hermann Weyl, *Symmetry* (Princeton: Princeton University Press, 1969)

Eugene Wigner, 'The Unreasonable Effectiveness of Mathematics in the Natural Sciences', *Communications in Pure and Applied Mathematics*, 13, (1960), pp. 222–337

Eugene Wigner, *Symmetries and Reflections* (Bloomington: Indiana University Press, 1967)

A. T. Winfree, *The Geometry of Biological Time* (New York: Springer-Verlag, 1980, corrected edn 1990)

Appendix 1

Equations for Icons

This appendix gives the formulas needed to draw the icons shown in chapter 9. You can iterate them by computer: the program follows the same lines as the one listed in appendix 2, but using the equations described below.

The equations take their simplest form in complex notation, $z = x + iy$. Writing $\bar{z}$ for the complex conjugate $x - iy$ and $Re(w)$ for the real part of a complex number w, then the equation for icons with symmetry group D_n is

$$f(z) = (a + bz\bar{z} + cRe(z^n))z + dz^{n-1}.$$

Here a, b, c, d are four arbitrary real numbers, although they have to be chosen with some care to get good results. For those unfamiliar with complex numbers, we give the formula in cartesian coordinates for small values of n.

When $n = 3$, the equation takes the form

$$f(x, y) = a\begin{pmatrix} x \\ y \end{pmatrix} + b\begin{pmatrix} x(x^2 + y^2) \\ y(x^2 + y^2) \end{pmatrix} + c\begin{pmatrix} x(x^3 - 3xy^2) \\ y(3x^2y - y^3) \end{pmatrix} + d\begin{pmatrix} x^2 - y^2 \\ -2xy \end{pmatrix}$$

When $n = 4$, the equation takes the form

$$f(x, y) = a\begin{pmatrix} x \\ y \end{pmatrix} + b\begin{pmatrix} x(x^2 + y^2) \\ y(x^2 + y^2) \end{pmatrix} + c\begin{pmatrix} x(x^4 - 6x^2y^2 + y^4) \\ y(4x^3y - 4xy^3) \end{pmatrix}$$

$$+ d\begin{pmatrix} x^3 - 3xy^2 \\ -3x^2y + y^3 \end{pmatrix}$$

When $n = 5$, the equation takes the form

$$f(x, y) = a\begin{pmatrix} x \\ y \end{pmatrix} + b\begin{pmatrix} x(x^2 + y^2) \\ y(x^2 + y^2) \end{pmatrix} + c\begin{pmatrix} x^5 - 10x^3y^2 + 5xy^4 \\ 5x^4y - 10x^2y^3 + y^5 \end{pmatrix}$$

$$+ d\begin{pmatrix} x^4 - 6x^2y^2 + y^4 \\ -4x^3y + 4xy^3 \end{pmatrix}$$

Here's a table of values of a, b, c, d that produce artistic icons when $n = 3$. We're not going to list any more, because we want you to experiment for yourself. Beware: many values cause the point being plotted to disappear rapidly to infinity; others lead to a few isolated dots. Be patient. For further instructions, consult *Symmetry in Chaos* by Field and Golubitsky, listed under Further Reading.

a	b	c	d
1.89	−1.10	0.17	−0.79
−1.89	1.80	0.00	1.34
1.00	−1.00	0.00	0.95
1.89	−1.10	0.21	0.60
1.35	−0.90	0.00	−0.80

Appendix 2

Computer Programs for Quilts

Here we describe two computer programs to draw quilt patterns: one for black-and-white, the other for colour. Because different computers use different graphics commands, the programs are written in pseudocode – code where you can guess what's intended. You should be able to convert them into programs that will run on your home computer. For the colour version to work you need high- resolution colour graphics.

The equations to be iterated are of the form:

$$f(x, y) = a\begin{pmatrix}\sin(2\pi x)\\ \sin(2\pi y)\end{pmatrix} + b\begin{pmatrix}\sin(2\pi x)\cos(2\pi y)\\ \sin(2\pi y)\cos(2\pi x)\end{pmatrix}$$

$$+ c\begin{pmatrix}\sin(4\pi x)\\ \sin(4\pi y)\end{pmatrix} + d\begin{pmatrix}\sin(6\pi x)\cos(4\pi y)\\ \sin(6\pi y)\cos(4\pi x)\end{pmatrix} + k\begin{pmatrix}x\\ y\end{pmatrix}$$

where a, b, c, d are four arbitrary real numbers and k is an arbitrary integer.

Black and white

input a, b, c, d, k

[*Comment*: Set initial values of x and y]
$x = 0.1{:}y = 0.3$
[*Comment*: First iterate to eliminate transients]
for $n = 1$ to 50
$x1 = a\sin(2\pi x) + b\sin(2\pi x)\cos(2\pi y) + c\sin(4\pi x) + d\sin(6\pi x)\cos(4\pi y) + kx$
$x1 = x1 - \text{int}(x1)$
[*Comment*: see note at end of programs]
$y1 = a\sin(2\pi y) + b\sin(2\pi y)\cos(2\pi x) + c\sin(4\pi y) + d\sin(6\pi y)\cos(4\pi x) + ky$
$y1 = y1 - \text{int}(y1)$
[*Comment*: see note at end of programs]

$x = x1{:}y = y1$
next n
[*Comment*: Procede with actual iteration and graphics]
for $n = 1$ to 1000
$x1 = a\sin(2\pi x) + b\sin(2\pi x)\cos(2\pi y) + c\sin(4\pi x) + d\sin(6\pi x)\cos(4\pi y) + kx$
$x1 = x1 - \text{int}(x1)$
[*Comment*: see note at end of programs]
$y1 = a\sin(2\pi y) + b\sin(2\pi y)\cos(2\pi x) + c\sin(4\pi y) + d\sin(6\pi y)\cos(4\pi x) + ky$
$y1 = y1 - \text{int}(y1)$
[*Comment*: see note at end of programs]
  $x = x1{:}y = y1$
  for $p = 0$ to 3
  for $q = 0$ to 3
  plot point with coordinates $(50x + 50p, 50y + 50q)$
[*Comment*: the $p - q$ loops draw an 3×3 array of copies of the same set, to make a tiling pattern. You may need to change the value 50 to make the picture fit nicely on your screen. You want to choose as large a value as possible to get the best resolution picture.]
  next q
  next p
  next n
  stop

Colour

dimension array A size 50×50
[*Comment*: You may need to change the value 50 to make the picture fit your screen. If so, change all other 50s below.]
  input a, b, c, d, k
  $x = 0.1{:}y = 0.3$
  for $n = 1$ to 20
  $x1 = a\sin(2\pi x) + b\sin(2\pi x)\cos(2\pi y) + c\sin(4\pi x) + d\sin(6\pi x)\cos(4\pi y) + kx$
  $x1 = x1 - \text{int}(x1)$
[*Comment*: see note at end of programs]
$y1 = a\sin(2\pi y) + b\sin(2\pi y)\cos(2\pi x) + c\sin(4\pi y) + d\sin(6\pi y)\cos(4\pi x) + ky$
$y1 = y1 - \text{int}(y1)$
[*Comment*: see note at end of programs]
$x = x1{:}y = y1$
next n
for $n = 1$ to 1000
$x1 = a\sin(2\pi x) + b\sin(2\pi x)\cos(2\pi y) + c\sin(4\pi x) + d\sin(6\pi x)\cos(4\pi y) + kx$
$x1 = x1 - \text{int}(x1)$
[*Comment*: see note at end of programs]
$y1 = a\sin(2\pi y) + b\sin(2\pi y)\cos(2\pi x) + c\sin(4\pi y) + d\sin(6\pi y)\cos(4\pi x) + ky$
$y1 = y1 - \text{int}(y1)$
[*Comment*: see note at end of programs]

```
x = x1:y = y1
A(50x, 50y) = A(50x, 50y) + 1
next n
for i = 1 to 50
for j = 1 to 50
for p = 0 to 3
for q = 0 to 3
plot point with coordinates (i + 50p, j + 50q) in a colour determined by the
value of A(i, j)
[Comment: For example, use red if A(i, j) ≤ 5, blue if 6 ≤ A(i, j) ≤ 10, etc.
The p − q loops draw an 8 × 4 array of copies of the same set, to make a tiling
pattern.]
   next q
   next p
   next j
   next i
   stop
```

Note on integer parts

On some computers the INT command in BASIC is the same as the mathematician's integer part, which is what we need here. On others, however, it works differently for negative numbers. To test, see whether your computer thinks that INT(−2.3) is −2 or −3. If it's −3, use INT. If it's −2, replace the command

```
x1 = x1 − int(x1)
```

by

```
if x ≥ 0 then x1 = x1 − int(x1) else x1 = x1 − int(x1) + 1
```

and similarly for $y1$.

Parameter values

To help you find interesting quilts, here's a table of parameter values that produce artistic results.

a	b	c	d	k
−0.59	0.20	0.10	0.00	0
0.25	−0.17	0.20	0.30	1
0.25	−0.30	0.20	0.17	−1
0.80	0.20	0.10	−0.10	2
0.46	0.00	0.00	−0.05	0
−0.46	0.00	0.00	−0.05	−2

Illustration Acknowledgements

Academic Press, London: Elwyn R. Berlekamp, John H. Conway and Richard K. Guy, *Winning Ways* – fig 10.6

Batchworth Press, London: *Larousse Encyclopaedia of Astronomy* – figs 6.1, 6.2

Librairie Classique Eugène Belin, Paris: *Pour La Science* – figs 1.9, 2.1, 3.6

Bohn, Scheltema and Holkema, Utrecht: Caroline H. Macgillavry, *Symmetry Aspects of* M. C. Escher's *Periodic Drawings* – figs 9.6, 9.10

Cambridge University Press, Cambridge: Paul Murdin and David Allen, *Catalogue of the Universe* – figs 6.4, 6.5; D'Arcy Thompson, *On Growth and Form* – fig 7.7

Chapman and Hall, London: R. McN. Alexander and G. Goldspink, *Mechanics and Energetics of Animal Locomotion* – figs 8.19, 8.21, 8.22

Jack Cohen – fig 7.10

Cordon Art BV, Baarn – figs 2.8, 9.6, 9.10

Fitzwilliam Museum, Cambridge: William Blake, *The Ancient of Days* – fig 1.1

W. H. Freeman, New York: Branko Grünbaum and G. C. Shephard, *Tilings and Patterns* – fig 4.13

W. N. Freeman, New York: Ilya Prigogine, *From Being to Becoming* (as fig 4.5, p. 189), fig 5.6

Brian Goodwin – fig 7.9

Haags Gemeentemuseum, s'Gravenhage – fig 2.8

Adam Hilger, Bristol: Marjorie Senechal, *Crystalline Symmetries – an Informal Mathematical Introduction* – fig 4.14

Hutchinson, London: A. C. Bishop, *An Outline of Crystal Morphology* – fig 4.18

George V. Kelvin, Science Graphics, Great Neck NY – fig 10.5

James Kilkelly – fig 6.11

Kuperard, London: P. P. Gambaryan, *How Mammals Run* – figs 8.1, 8.2, 8.3, 8.4, 8.5, 8.6, 8.7, 8.12, 8.13

Larousse, Paris: *The Larousse Encyclopaedia of Astronomy* – figs 6.1, 6.2

Longman Group UK, Harlow: F. C. Phillips, *An Introduction to Crystallography* – figs 4.1, 4.2, 4.6, 4.8, 4.9

Macmillan Magazines Ltd, London: *Nature* – figs 6.8, 8.18

McGraw Hill Book Co (UK) Ltd, Maidenhead: M. S. El Naschie, *Stress, Stability, and Chaos in Structural Engineering: an Energy Approach* – fig 1.5

Dr G. T. Meaden/Fortean Picture Library, *Crop Circles, Exton, Hampshire, 1990* – fig 1.12

MIT Press, Cambridge MA: Whitman Richards (ed.), *Natural Computation* – fig 3.5

National Academy Press, Washington DC: Lynn Arthur Steen (ed.), *On the Shoulders of Giants* – figs 3.1, 7.3, 7.14

New Scientist, London – fig 6.10

Odhams Press, London: *The Marvels and Mysteries of Science* – fig 6.3

Oxford University Press, Oxford: Edwin Hubble, *The Realm of the Nebulae* – fig 6.6

R. Peper for the Prigogine – fig 5.6

Pergamon Press, Oxford: Jack Cohen, *Living Embryos* – figs 7.4, 7.5, 7.6, 7.11

Princeton University Press, Princeton: James Binney and Scott Tremaine, *Galactic Dynamics* – figs 6.7; John Tyler Bonner, *Morphogenesis* – figs 7.8, 7.12; Vera C. Rubin and George V. Coyne, S. J. (eds), *Large Scale Motions in the Universe* – fig 6.11; Hermann Weyl, *Symmetry* – fig 9.7

Science Photo Library, London – fig 7.2

Scientific American, New York – figs 1.10, 6.9, 7.14, 7.15, 10.5

Springer-Verlag, New York: Martin Golubitsky, Ian Stewart and David G. Schaeffer, *Singularities and Groups in Bifurcation Theory vol.2* – figs 5.7, 5.8; *Mathematical Intelligencer* – fig 9.11

Harry Swinney: A. Brandstater and H. L. Swinney, *Physical Review* A 35 2207–2220 (1987) – fig 5.2

United Media, New York: J. Davis, *Garfield: Bigger than Life* – fig 8.8

University of California Press, Berkeley: John G. Burke, *Origins of the Science of Crystals* – fig 4.5

James C. Walters – fig 1.10

John Wiley & Sons Ltd, Chichester: *Dana's Manual of Mineralogy* – figs 4.12, 4.15; J. M. T. Thompson, *Instabilities and Catastrophes in Science and Engineering* – fig 1.6; J. M. T. Thompson and H. B. Stewart, *Nonlinear Dynamics and Chaos* – fig 9.1.

Index

READ MORE IN PENGUIN

READ MORE IN PENGUIN 

SCIENCE AND MATHEMATICS

QED Richard Feynman
The Strange Theory of Light and Matter

'Physics Nobelist Feynman simply cannot help being original. In this quirky, fascinating book, he explains to laymen the quantum theory of light – a theory to which he made decisive contributions' – *New Yorker*

Does God Play Dice? Ian Stewart
The New Mathematics of Chaos

To cope with the truth of a chaotic world, pioneering mathematicians have developed chaos theory. *Does God Play Dice?* makes accessible the basic principles and many practical applications of one of the most extraordinary – and mind-bending – breakthroughs in recent years.

Bully for Brontosaurus Stephen Jay Gould

'He fossicks through history, here and there picking up a bone, an imprint, a fossil dropping and, from these, tries to reconstruct the past afresh in all its messy ambiguity. It's the droppings that provide the freshness: he's as likely to quote from Mark Twain or Joe DiMaggio as from Lamarck or Lavoisier' – *Guardian*

The Blind Watchmaker Richard Dawkins

'An enchantingly witty and persuasive neo-Darwinist attack on the anti-evolutionists, pleasurably intelligible to the scientifically illiterate' – Hermione Lee in the *Observer* Books of the Year

The Making of the Atomic Bomb Richard Rhodes

'Rhodes handles his rich trove of material with the skill of a master novelist … his portraits of the leading figures are three-dimensional and penetrating … the sheer momentum of the narrative is breathtaking … a book to read and to read again' – Walter C. Patterson in the *Guardian*

Asimov's New Guide to Science Isaac Asimov

A classic work brought up to date – far and away the best one-volume survey of all the physical and biological sciences.

READ MORE IN PENGUIN

SCIENCE AND MATHEMATICS

The Panda's Thumb Stephen Jay Gould

More reflections on natural history from the author of *Ever Since Darwin*. 'A quirky and provocative exploration of the nature of evolution ... wonderfully entertaining' – *Sunday Telegraph*

Einstein's Universe Nigel Calder

'A valuable contribution to the demystification of relativity' – *Nature*. 'A must' – *Irish Times*. 'Consistently illuminating' – *Evening Standard*

Gödel, Escher, Bach: An Eternal Golden Braid
Douglas F. Hofstadter

'Every few decades an unknown author brings out a book of such depth, clarity, range, wit, beauty and originality that it is recognized at once as a major literary event' – Martin Gardner. 'Leaves you feeling you have had a first-class workout in the best mental gymnasium in town' – *New Statesman*

The Double Helix James D. Watson

Watson's vivid and outspoken account of how he and Crick discovered the structure of DNA (and won themselves a Nobel Prize) – one of the greatest scientific achievements of the century.

The Quantum World J. C. Polkinghorne

Quantum mechanics has revolutionized our views about the structure of the physical world – yet after more than fifty years it remains controversial. This 'delightful book' (*The Times Educational Supplement*) succeeds superbly in rendering an important and complex debate both clear and fascinating.

Mathematical Circus Martin Gardner

A mind-bending collection of puzzles and paradoxes, games and diversions from the undisputed master of recreational mathematics.

READ MORE IN PENGUIN

SCIENCE AND MATHEMATICS

The Dying Universe Paul Davies

In this enthralling book the author of *God and the New Physics* tells how, from the instant of its fiery origin in a big bang, the universe has been running down. With clarity and panache Paul Davies introduces the reader to a mind-boggling array of cosmic exotica to help chart the cosmic apocalypse.

The Newtonian Casino Thomas A. Bass

'The story's appeal lies in its romantic obsessions ... Post-hippie computer freaks develop a system to beat the System, and take on Las Vegas to heroic and thrilling effect' – *The Times*

Wonderful Life Stephen Jay Gould

'He weaves together three extraordinary themes – one palaeontological, one human, one theoretical and historical – as he discusses the discovery of the Burgess Shale, with its amazing, wonderfully preserved fossils – a time-capsule of the early Cambrian seas' – *Mail on Sunday*

The New Scientist Guide to Chaos edited by Nina Hall

In this collection of incisive reports, acknowledged experts such as Ian Stewart, Robert May and Benoit Mandelbrot draw on the latest research to explain the roots of chaos in modern mathematics and physics.

Innumeracy John Allen Paulos

'An engaging compilation of anecdotes and observations about those circumstances in which a very simple piece of mathematical insight can save an awful lot of futility' – Ian Stewart in *The Times Educational Supplement*

Fractals Hans Lauwerier

The extraordinary visual beauty of fractal images and their applications in chaos theory have made these endlessly repeating geometric figures widely familiar. This invaluable new book makes clear the basic mathematics of fractals; it will also teach people with computers how to make fractals themselves.

READ MORE IN PENGUIN

HISTORY

The World Since 1945 T. E. Vadney
New edition

From the origins of the post-war world to the collapse of the Soviet Bloc in the late 1980s, this masterly book offers an authoritative yet highly readable one-volume account.

Ecstasies Carlo Ginzburg

This dazzling work of historical detection excavates the essential truth about the witches' Sabbath. 'Ginzburg's learning is prodigious and his journey through two thousand years of Eurasian folklore a *tour de force*' – *Observer*

The Nuremberg Raid Martin Middlebrook

'The best book, whether documentary or fictional, yet written about Bomber Command' – *Economist*. 'Martin Middlebrook's skill at description and reporting lift this book above the many memories that were written shortly after the war' – *The Times*

A History of Christianity Paul Johnson

'Masterly … It is a huge and crowded canvas – a tremendous theme running through twenty centuries of history – a cosmic soap opera involving kings and beggars, philosophers and crackpots, scholars and illiterate exaltés, popes and pilgrims and wild anchorites in the wilderness'– Malcolm Muggeridge

The Penguin History of Greece A. R. Burn

Readable, erudite, enthusiastic and balanced, this one-volume history of Hellas sweeps the reader along from the days of Mycenae and the splendours of Athens to the conquests of Alexander and the final dark decades.

Modern Ireland 1600–1972 R. F. Foster

'Takes its place with the finest historical writing of the twentieth century, whether about Ireland or anywhere else' – Conor Cruise O'Brien in the *Sunday Times*

READ MORE IN PENGUIN

HISTORY

The Guillotine and the Terror Daniel Arasse

'A brilliant and imaginative account of the punitive mentality of the revolution that restores to its cultural history its most forbidding and powerful symbol' – Simon Schama.

The Second World War A J P Taylor

A brilliant and detailed illustrated history, enlivened by all Professor Taylor's customary iconoclasm and wit.

Daily Life in Ancient Rome Jerome Carcopino

This classic study, which includes a bibliography and notes by Professor Rowell, describes the streets, houses and multi-storeyed apartments of the city of over a million inhabitants, the social classes from senators to slaves, and the Roman family and the position of women, causing *The Times Literary Supplement* to hail it as a 'thorough, lively and readable book'.

The Anglo-Saxons Edited by James Campbell

'For anyone who wishes to understand the broad sweep of English history, Anglo-Saxon society is an important and fascinating subject. And Campbell's is an important and fascinating book. It is also a finely produced and, at times, a very beautiful book' – *London Review of Books*

The Making of the English Working Class E. P. Thompson

Probably the most imaginative – and the most famous – post-war work of English social history. 'A magnificent, lucid, angry historian ... E. P. Thompson has performed a revolution of historical perspective' – *The Times*

The Habsburg Monarchy 1809 –1918 A J P Taylor

Dissolved in 1918, the Habsburg Empire 'had a unique character, out of time and out of place'. Scholarly and vividly accessible, this 'very good book indeed' (*Spectator*) elucidates the problems always inherent in the attempt to give peace, stability and a common loyalty to a heterogeneous population.

READ MORE IN PENGUIN

HISTORY

Citizens Simon Schama

The award-winning chronicle of the French Revolution. 'The most marvellous book I have read about the French Revolution in the last fifty years' – Richard Cobb in *The Times*

To the Finland Station Edmund Wilson

In this authoritative work Edmund Wilson, considered by many to be America's greatest twentieth-century critic, turns his attention to Europe's revolutionary traditions, tracing the roots of nationalism, socialism and Marxism as these movements spread across the Continent creating unrest, revolt and widespread social change.

Jasmin's Witch Emmanuel Le Roy Ladurie

An investigation into witchcraft and magic in south-west France during the seventeenth century – a masterpiece of historical detective work by the bestselling author of Montaillou.

Stalin Isaac Deutscher

'The Greatest Genius in History' and the 'Life-Giving Force of socialism'? Or a despot more ruthless than Ivan the Terrible and a revolutionary whose policies facilitated the rise of Nazism? An outstanding biographical study of a revolutionary despot by a great historian.

Aspects of Antiquity M. I. Finley

Profesor M. I. Finley was one of the century's greatest ancient historians; he was also a master of the brief, provocative essay on classical themes. 'He writes with the unmistakable enthusiasm of a man who genuinely wants to communicate his own excitement' – Philip Toynbee in the *Observer*

British Society 1914–1945 John Stevenson

'A major contribution to the *Penguin Social History of Britain*, which will undoubtedly be the standard work for students of modern Britain for many years to come' – *The Times Educational Supplement*

READ MORE IN PENGUIN

ARCHAEOLOGY

Missing Links John Reader

In this classic account John Reader tells the story of the hunt for the fossil remains of our earliest ancestors – a story enlivened by fraud, controversy and more than its fair share of English, American and European eccentrics and enthusiasts. 'Fascinating and carefully documented' – *Observer*

The Scars of Evolution Elaine Morgan

'Elaine Morgan seems to have succeeded where the professionals have failed. She has made a genuine contribution to evolutionary theory which synthesizes research from a wide range of scientific disciplines; and she has presented it in a form which is accessible to the interested lay reader ... remarkable ... an exceptionally well-written book' – *British Medical Journal*

The Pyramids of Egypt I. E. S. Edwards
Revised Edition

Dr Edwards offers us the definitive work on these gigantic tombs, drawing both on his own original research and on the work of the many archaeologists who have dug in Egypt. This revised edition includes recent discoveries and research.

Lucy's Child Donald Johanson and James Shreeve

'Superb adventure ... *Lucy's Child* burns with the infectious excitement of hominid fever ... the tedium and the doubting, and the ultimate triumph of an expedition that unearths something wonderful about the origins of humanity' – *Chicago Tribune*

Dawn of a Millennium Erich Harth

With our 'Stone Age bodies and Stone Age brains', claims Erich Harth in this profound and disturbing analysis of human evolution, we now 'face tasks for which we are genetically unprepared'. 'A fascinating work ... This book should make us sit up and think – and act' – Roger Penrose

READ MORE IN PENGUIN

RELIGION

The Gnostic Gospels Elaine Pagels

In a book that is as exciting as it is scholarly, Elaine Pagels examines these ancient texts and the questions they pose and shows why Gnosticism was eventually stamped out by the increasingly organized and institutionalized Orthodox Church. 'Fascinating' – *The Times*

Islam in the World Malise Ruthven

This informed and informative book places the contemporary Islamic revival in context, providing a fascinating introduction – the first of its kind – to Islamic origins, beliefs, history, geography, politics and society.

The Orthodox Church Timothy Ware

In response to increasing interest among western Christians, and believing that a thorough understanding of Orthodoxy is necessary if the Roman Catholic and Protestant Churches are to be reunited, Timothy Ware explains Orthodox views on a vast range of matters from Free Will to the Papacy.

Judaism Isidore Epstein

The comprehensive account of Judaism as a religion and as a distinctive way of life, presented against a background of 4,000 years of Jewish history.

Mysticism F. C. Happold

What is mysticism? This simple and illuminating book combines a study of mysticism – as experience, as spiritual knowledge and as a way of life – with an illustrative anthology of mystical writings, ranging from Plato and Plotinus to Dante.

Eunuchs for Heaven Uta Ranke-Heinemann

'No other book on the Catholic moral heritage unearths as many spiteful statements about women ... it is sure to become a treasure-chest for feminists ... Uta Ranke-Heinemann's research is dazzling' – Jason Berry in *The New York Times*

READ MORE IN PENGUIN 

PHILOSOPHY

What Philosophy Is Anthony O'Hear

'Argument after argument is represented, including most of the favourites … its tidy and competent construction, as well as its straightforward style, mean that it will serve well anyone with a serious interest in philosophy' – *The Journal of Applied Philosophy*

Montaigne and Melancholy M. A. Screech

'A sensitive probe into how Montaigne resolved for himself the age-old ambiguities of melancholia and, in doing so, spoke of what he called the "human condition"' – Roy Porter in the *London Review of Books*

Labyrinths of Reason William Poundstone

'The world and what is in it, even what people say to you, will not seem the same after plunging into *Labyrinths of Reason* … Poundstone's book merits the description of *tour de force*. He holds up the deepest philosophical questions for scrutiny and examines their relation to reality in a way that irresistibly sweeps readers on' – *New Scientist*

I: The Philosophy and Psychology of Personal Identity
Jonathan Glover

From cases of split brains and multiple personalities to the importance of memory and recognition by others, the author of *Causing Death and Saving Lives* tackles the vexed questions of personal identity.

Ethics Inventing Right and Wrong J. L. Mackie

Widely used as a text, Mackie's complete and clear treatise on moral theory deals with the status and content of ethics, sketches a practical moral system, and examines the frontiers at which ethics touches psychology, theology, law and politics.

The Central Questions of Philosophy A. J. Ayer

'He writes lucidly and has a teacher's instinct for the helpful pause and reiteration … an admirable introduction to the ways in which philosophic issues are experienced and analysed in current Anglo-American academic milieux' – *Sunday Times*